The Landships of Lincoln

SECOND EDITION

The story of Lincoln's part in the creation of the worlds first fighting tanks

All the best,
Richard Pullen

Richard Pullen

First Published in Great Britain in 2007 by TUCANN books

Published & Produced by: TUCANN*design&print*, 19 High Street, Heighington Lincoln LN4 1RG
Tel & Fax: 01522 790009
www.tucann.co.uk

CONTENTS

INTRODUCTION 5

CHAPTER I
Tritton, Wilson and the creation of an alternative 7

CHAPTER II
Tank Production 41

CHAPTER III
Medium Tanks 72

CHAPTER IV
The Heavy Tanks 84

CHAPTER V
For Valour 96

CHAPTER VI
Bank in the Tank 106

CHAPTER VII
A tank in your Town 122

CHAPTER VIII
Where are they now? 129

APPENDIX 1
Taped reminiscences of William Foster and Co Ltd employees
Undertaken by the 'Lincoln Tank Group' between 1977 and 1984...........151

APPENDIX II
Drawings of Great War military vehicles
produced in Lincolnshire's factories 164

BIBLIOGRAPHY
Recommended reading and acknowledgements 186

With thanks to individuals, museums and agencies too numerous to mention, but including the invaluable assistance freely given by Ray Hooley, Philippe Gorczynski of the Association Tank de Flesquieres, David Fletcher and the Bovington Tank Museum, Bill and Charlotte Baker. Lastly, but just as vital was the help and encouragement given by Joanne, Grace and Avro the dog, without whom I probably never would have started, let alone finished.

This book is dedicated to those who designed and built the first tanks, to those who fought and died in them and to those who now strive to preserve their memory and our heritage

About the Author

Richard Pullen has lived in Lincoln for most of his life and has strong family links with the creation of the tank in Lincoln. His Grandfather worked at William Foster and Co Ltd on tank production during the First World War. Richards great-uncle Tom worked at Ruston and Hornsbys in Lincoln during the Second World War, his speciality being welding the turrets on Matilda tanks. Richards late father, Bob, spent 30 years of his working life at Ruston Gas Turbines in the city and took a particular interest in the restoration of the Cambrai veteran Mk IV female Flirt II that is now on display in the Museum of Lincolnshire Life on Burton Road. Richard himself spent several months as a volunteer at the museum showing visitors and school parties around the restored tank and was one of the founder members of 'Friends of the Lincoln Tank' a newly formed group promoting Lincoln's ties with the first tanks.

Richard has been a professional field archaeologist for over a decade, gaining his masters degree in archaeological heritage management in 2002. Although his particular interest is industrial archaeology, during his archaeological career he has directed sites of all periods throughout the British Isles.

INTRODUCTION

Throughout the history of warfare one of the main strengths of any army has been its mobility. An army needs the ability to strike whenever and wherever they are needed, but with the outbreak of war in 1914 the worlds fighting forces soon became embroiled in a terrible, new kind of fighting known as trench warfare. Suddenly, mobility was a thing of the past and thousands of fighting men from both sides were dying every day just to maintain a stalemate in 450 miles (725 Kilometres) of fortified trenches running from the northern coast of Belgium down to the Italian Alps. Since the start of the industrial revolution in the 18th century, weapons technology had progressed steadily, but the beginning of the twentieth century saw an unprecedented advancement in the arms race. In early 1915 David Lloyd George could already see how the outcome of the new war would be decided when he said *'this is an engineers war'*. Mechanization of the battlefield had arrived and weapons such as the aeroplane, poison gas, Dreadnaughts, the armoured car, the submarine, flame-throwers and pinpoint accurate field artillery were all terrifying new inventions to be put to use in this new kind of war. However, a combination of two main inventions can be said to define the horror of the trenches, they were barbed wire and the machine gun. A team of three well-trained men and one machine gun could now do what a few years before would have taken hundreds of riflemen. The deadly accuracy and awesome power of the machine gun created the need for armies to 'dig in' and impenetrable belts of barbed wire between the trenches kept them there. The ability for either side to advance and fight had been all but lost and whichever side was the first to breach the wire and combine mechanised firepower with their army's lost mobility would break the deadlock and perhaps even win the war.

It was all supposed to be over by Christmas 1914, but Christmas came and went and still the unimaginable horror of trench warfare raged on, an entire generation of the worlds young men was being pointlessly sacrificed in futile battles in France, Belgium, Turkey and Palestine. The infantry was ordered to charge the enemy in textbook 17th century style, but their foe was armed with the most terrible killing technology the 20th century had to offer. A fixed bayonet or a cavalry charge would be no match for trench mortars and machine guns. When confronted with these images of hell on earth in was inevitable that inventive minds on both sides and from all ranks would begin to consider alternatives and in England a small group of men began to have the glimmering of a remarkable idea. It was to be an idea that would lead to a small agricultural engineering company in Lincoln being asked to create the tank, a new mobile weapon that could travel over trenches, shell holes, and barbed wire and eventually change the face of war forever. From its first, faltering introduction into the trenches of 1916 to the most recent conflicts in Afghanistan and Iraq no army would consider going into war without tanks. From the moment, over 90 years ago, that the secrecy was lifted and the tank was unveiled it became much more than just a weapon; it became a potent demonstration of military and political power, a show of a nations fighting strength and a deterrent to potential enemies.

Over the years hundreds of books have been written covering the apocalyptic destruction of the Great War, and the tank is a terrifying and inhuman part of that terrible conflict. However, this book was not written to describe the horror into which the early tanks and their ill-prepared crews were thrown. It was written in an attempt to cover the production and use of the first fighting tanks from the perspective of the City of Lincoln and the local people who built them. This is not so much a story of the First World War as a story of a remarkable invention and the creative minds and tenacious men and women who created it.

CHAPTER I

Tritton, Wilson and the creation of an alternative

Before any invention can be a success, three main requirements need to be in place and aligned. There must be a social requirement for the new creation, there must be a political need and perhaps most obvious of the three, the technology must exist in order to ouild it. By 1916, after two years of deadlock, the social and political factors were very much in place and although some of the technologies were still very much in their infancies, they were advanced enough for inspired people to begin to think seriously about the creation of a terrifying new weapon, a weapon that may even turn the tide of the war.

When one asks who invented the tank there is an seemingly never ending list of names such as Ernest Swinton, Lancelot de Mole, Murray Sueter, Thomas Hetherington, Winston Churchill, Tennyson d'Eyncourt, John *'Boney'* Fuller, A.C. Nesfield, Bramah Diplock, Albert Stern, Sir Arthur Conan Doyal, H.G. Wells and many more besides. It is true that many of these men played their own invaluable part in the protracted trail that led to the first tank and almost every author who has written on the subject seems to have championed the cause of a different contender. However, the undeniable fact is that these people had only imagined or considered the tank in theory, for instance Swinton suggested the idea of armoured trench crossers in 1914 and H.G. Wells predicted *'Land Ironclads'* early in the Boar War. However, it is one thing to have a revolutionary concept and quite another to overcome all design and engineering hurdles and put that concept into practice. If I were to use this page to describe my vision for a hover car or a time machine, would that mean that I had invented these things, of course not. Invention is much more than vague ideas in the dusty recesses of someone's mind or rough sketches at the footnote on a letter, it entails detailed design drawings, precise engineering, metallurgy, economic factors, feasibility studies and endless testing to name but a few. Whoever had the original idea for the tank is now lost in the mists of time, and in a way hardly matters at all. The central fact is that from a list of hundreds of contenders, just two men were able to claim that they did not just sit around in the war office having pointless arguments about who thought of the idea, they actually did something about it and became responsible for overcoming the engineering and design problems to physically create the fearsome new weapon. Those two men were Major Walter Gordon Wilson and Sir William Ashbee Tritton. A Royal Commission on awards to inventors was convened in October 1919 to decide if anyone in particular deserved special mention or financial reward for the creation of the tank. They concluded that, although several people such as Hetherington, Stern and Swinton deserved special mention for the part they had played, by far the largest share of any reward for inventing the tank should go to Wilson and Tritton for their work in Lincoln and they were awarded £15,000 to be shared equally between them. William Tritton gave most of his share of the money to his employees as he felt that he had done well enough by being granted a Knighthood in 1917 and he had also enjoyed national fame and increased profits from War Office contracts since the beginning of hostilities. Tritton told the Royal Commission that he had not made his fortune from the tanks; in fact he had received no more than 33 shillings per tank produced. Although this may not sound like very much and actually only equates to just over £1.50 per machine in today's terms, it must be remembered that at the time a private risking his life in the Tank Corps was only getting 2s 6d per day, just 12 pence today. In return for sharing the Royal Commission award a complimentary dinner in honour of Tritton and Wilson was organized by Fosters technical and commercial staff. The dinner was held at the White Hart Hotel in Lincoln on the evening of the 5th December 1919 '*in order to celebrate their official recognition as the originators and designers of the fighting tank'*. Celebrations were well founded as prior to the start of tank production in 1915 Fosters had an impressive sales figure for the year of £187,000. However, just one year later sales had topped an unbelievable £400,000, figures that Fosters Chairman of the Directors, C.W. Pennall, described at the time as '*speaking for themselves*'.

Walter Wilson was born in 1874 and had started out in a military fashion, being educated as a Navel cadet at Dartmouth and Kings College, Cambridge. When he left he had a first class honours degree in mechanical sciences and a fertile mind full of ideas. It is no overstatement to say that he was an engineering and design genius. He designed and built his own cars, completely redesigned the sailing boat and even dabbled with early aircraft. All of the machines he worked on were incredibly technologically advanced and of very high quality. Wilson had a long tradition of military invention behind him, having designed a combined armoured car and gun tractor for Armstrong Whitworth in 1906, unfortunately the thing was ahead of it's time and they could not persuade anyone to buy it. Wilson left Armstrongs in 1908 and became a freelance engineer, but in 1914 war was declared and he heeded his countries call and joined the newly formed Royal Navy Armoured Car Service. It was perhaps inevitable that as he was an officer in the Armoured Car Service and also a first class engineer, he would eventually become involved with the Admiralties Landships Committee and the creation of the first tanks. The story of Wilsons involvement with tanks really began in 1915 when he was seconded to work at the small Lincoln based engineering firm of William Foster and Co Ltd. Whilst there he met and became firm friends with the company's Managing Director, William Tritton. Fosters had been awarded the contract to develop designs for various trench-crossing machines and Wilson and Tritton became a dynamic design team who between them would eventually create the first track laying armoured fighting vehicle.

Design genius and co-creator of the tank, Walter Wilson

Walter Wilson used his share of the Royal Commission award to continue in engineering after the war, working on a number of projects, but specializing in transmission units. In 1928 he designed and created the pre-selector gearbox and soon after started the company of Improved Gears Ltd in order to develop and promote his invention. Once the new gear system had been perfected Wilson changed the name of the company to Self Changing Gears Ltd. The company thrived, working on many and varied projects including cars, boats, trains and during the second world war, they even worked on tanks again. Walter Wilson died in 1957 at the age of 83 and two years later Self Changing Gears Ltd was taken over by Leyland Motors. Leyland kept the name until 1986 when they sold all rights to the Cummins group. Walter Wilson was always proud of his pioneering work on the tanks and their transmission units in particular. Many years after the Great War when gearbox problems reared their heads again on the then new Chieftain tanks Wilson had just grumbled that '*this could not have happened if matters had been conducted as during World War I'*

Walter Wilson was destined to become a military engineer from the moment he had been born whereas his tank design partner William Tritton started as an agricultural engineer and would have quite happily continued as such if it had not been for the war. Tritton was born in Islington in 1875 to William and Ellen Tritton. Young Williams farther was a successful London stockbroker and he was determined to ensure that his son would have a first class education. William Tritton Jnr. was educated at Christs College, Finchley

William Tritton in a typically defiant pose (The Tank Museum, Bovington)

and Kings College, London. He did well at school and when he left in 1891 he began his apprenticeship at Gwynns Pumps of Hammersmith. In his early work for Gwynns he showed a flair for design and a tenacious, imaginative character, which would serve him well throughout his career. During a strike at Gwynns he was promoted to shop foreman, but resolved to leave the firm as soon as he could after an incident during which he was docked four pence from his wages for tipping a porter. Tritton eventually left Gwynns and spent the last few years of the 19th century moving through an impressive list of some of Englands finest engineering companies including Thornycroft and Garratt. It was from Garratts that in 1905 Tritton was head hunted by Fosters Chairman W.T. Page. The small Lincoln based company had been founded by Poterhanworth born William Foster in 1846 as a flourmill on Lincolns Waterside North. In the early 1850s Foster transferred all machinery in the mill from water over to the more reliable new steam power. He was so impressed with the transformation that in 1856 he took the remarkable decision of abandoning the milling business he had practiced for 10 years in order to form an engineering foundry instead. Surprisingly, considering the amount of established engineering companies in and around Lincoln, his gamble paid off and the newly formed 'Wellington Foundry' flourished. Fosters were soon producing threshing machines and portable engines that would become world renowned for their high quality and workmanship. William Foster was not only a man with a strong entrepreneurial talent he was also a man with a strong social and political conscience, who believed he could do great things for the people of Lincoln. He started in a small way in 1852 by standing as the Liberal candidate for Lincoln and was returned as member for Middle Ward. Foster continued to spend all of his spare time working within Local Government until in 1864 he was appointed Mayor of Lincoln. Although William Foster would only be Mayor for a single year, he certainly made the most of his time in office. He served on almost every committee possible including bridges, markets, lighting and even drainage and sewerage. During his time as Lincolns Mayor Foster became a popular and well respected figure around the city and his time in office was commemorated on several new developments and civic improvements such as the water hydrant outside St. Peter in Gowts Church on the High Street which still bears his name to this day. Perhaps it is mere coincidence that the year after sitting on the committee responsible for street lighting William Fosters Wellington foundry received an order from Lincoln City Council for a batch of gas-lamp standards to be placed all around the upper city. Most were usurped by 1960s concrete replacements, which have in turn recently been replaced by modern Victorian styled cast iron ones, but at least three of the original Foster standards still survive in their original positions. One can be seen leaning at an alarming angle half way up Steep Hill, a second still stands outside the Castle Hotel

William Foster 1816 -1876.

The William Foster water hydrant outside St. Peter at Gowts Church on Lincoln High Street

off the Bailgate and the third is in the grounds of Lincoln's magnificent 10th century Cathedral. The Foster lamp-post outside the Castle hotel recently disappeared and was replaced by a modern galvanized example. Local history groups were horrified until it was made clear that the lamp-post was merely being restored and after several weeks it was returned to its former place in as-new condition.

Unfortunately, William Foster, the companies founder and one of Lincoln's most popular and well known civic figures, contracted typhoid and died on 14th October 1876 at the age of just 60 years old. He was buried in the quiet and almost forgotten Eastgate Cemetery, where his grave still exists along with that of his wife Mary and several other family members. William Foster is in illustrious company in Eastgate Cemetery surrounded by the graves bearing the names of other Lincoln notables including Rainforth, Duckering, Ruddock, Scorer and Ruston. The company of William Foster and Co LTD continued to trade after their founders death and were happy to spend the next thirty years creating a range of high quality agricultural machines for the world market, but despite moving to larger premises on Firth Road in 1899 they remained a modest company and never expanded into the large mass production techniques practiced by the likes of Rustons, Aveling-Barford, Garrets or Foden.

In 1905 William Tritton accepted Pages offer to move to Fosters and take over as General Manager. He moved to Lincoln, where he would continue to live at Cliff Dene on Eastcliffe Street for 41 years until his death in September 1946 at the age of 71. His body was cremated at Wilford Hill Cemetery in Nottingham, he left instructions that there should be no ceremony, beyond that necessary, no flowers and that no mourners should attend, not even his devoted wife Isabel.

When Tritton had first moved to William Fosters the company's financial position looked bleak. However, the challenge suited him perfectly and he was determined to do all he could to help the company to recover its former glory. To this end he rarely entered his own office, instead he worked very closely with the drawing and design team and was personally responsible for the redesign of many of Fosters existing machines. He also approved a surprising amount of completely new designs for production, many of which were specifically for export to the farthest reaches of the empire. Their work during the war would benefit the company greatly and they started a programme of expansion. In 1927 Fosters bought out the struggling Gwynns Pumps, making William Tritton probably the only apprentice ever to become managing director of his old firm.

William Rigby, draughtsman and tank designer

Neither Tritton nor Wilson were the sort of men who suffered fools gladly, but the combination of Trittons experience, tenacious character and technical knowledge, coupled with Walter Wilsons design and engineering genius made them a formidable team. These qualities were to serve them well when they were charged with the creation of the first tracked armoured fighting vehicles in 1915, but their work could not have taken place without a third man with whom Tritton and Wilson worked very closely during these early days. William Rigby was a gifted draughtsman with a keen grasp of engineering principals and a knack for being able to push the available technology to it's limits. Rigby went to school in Lincoln and in 1904 he became an apprentice at Fosters, quickly rising through the ranks until he had gone from drawing shop apprentice to become the companies chief draughtsman. His talent for producing highly detailed engineering drawings in an incredibly short time was a great asset to Fosters and he was directly responsible for creating the first drawings from which the tanks were built. He was a dedicated workaholic and it was often said around the factory that Rigbys drawing board actually folded out into a bed so that he never had to go home. Despite Rigbys dedication and his wartime achievements in tank design he was only too glad when the armistice was declared and the company could return to its pre-war role.

He continued to work at Fosters until he retired to his home on Lincolns Wragby Road in 1963. Right up until his retirement he was responsible for designing many new products including several special one-off orders for massive water pumps. William Rigby continued to give lectures on technical drawing, hydrodynamics and tank production right up until his death in January 1982.

Although Tritton, Wilson and Rigby can be credited with creating the first tank the ancestry of armoured fighting vehicles goes back much further in time than their involvement during the First World War. It can be traced back to the very first time that one army had a military advantage over another, thus creating the need for a fighting force to attack their enemy at close quarters from the relative safety of a mobile strong point. Asking who first had the idea of the tank is like asking who first thought of the wheel and seems futile. The idea of mobile armoured strong-points goes back at least 3000 years to when the Assyrians had war chariots covered with a protective canopy of toughened leather. Since then, countless others have created armoured fighting vehicle concepts and sketches such as Leonardo Da Vinci, who presented his patron with an inspired, but characteristically impracticable design for a mobile fort in the 15th century. With the advent of reliable portable steam engines in the 1800s various forward thinking individuals and companies such as James Cowen and Fowlers of Leeds, to name but two, attempted to create a steam powered armoured fighting vehicle. Cowens plan, like so many before and since was rejected out of hand by the war office, but each time a new armoured fighting vehicle was proposed or created another link was forged on the technological chain, which would eventually lead to the creation of the tank. Most of the inventions preceding the First World War had more in common with the emerging armoured car than with the tank, as we know it today. Armoured cars had already proved useless on the near liquid mud and barbed wire strewn morass of Ypres and would never be able to play a decisive part in breaking the trenches. The artillery barrages of the Great War had created an impassable featureless sea of mud and shell holes. The lowlands of Belgium are very much like the Fens of Lincolnshire, both areas of reclaimed land only just above sea level, where a delicate balance between land and sea is maintained by a system of dykes and drains. Belgiums network of dykes had been obliterated by two years of shelling leading to a dramatic rise in the water table and the creation of the horrific landscape that is conjured up in the minds eye whenever the trenches are mentioned. Therefore, the defining characteristic of the tank and also the main difference between it and everything that had come before it would be its ability to cross rough, or disturbed ground by the use of caterpillar tracks.

Many people still maintain that the concept of track laying vehicles was actually first thought of in America at some time in the late 1800s, but in fact a track related patent was granted to Richard Lovell Edgeworth in London as early as 1770. However, gaining a patent for his concept of *'an endless track running over wheels'* appears to be as far as Mr. Edgeworth ever got, as no physical manifestation of his tracks is recorded. It was not until more than one hundred years after this inspiration that the American track system was created when in late 1880 Mr. Batter introduced a tracked steam tractor. Mr. Batter is unlikely to have known about Lovell Edgeworths patent and his new creation was inspired. It was perfectly suited to the vast open cornfields and unbroken land of the prairies and sold in great numbers, but in Europe with its small, enclosed patchwork fields, very few farmers or manufacturers could see the value of such a radical vehicle. Richard Hornsbys and Sons of Grantham in Lincolnshire was one of the few manufacturers to try their hand in the tracked vehicle markets of Europe. Hornsbys had become world renowned as quality steam engine makers, but by the beginning of the twentieth century they were concentrating on the future in the form of petrol and oil engined machines. It is difficult nowadays to grasp how unfamiliar and misunderstood the concept of oil fired engines was at the start of the 20th century. At the time steam was still King, and even by 1914 the entire British Army had a total of only 254 motor vehicles, including lorries, cars and motorcycles. In 1909 the Automobile Association had organised a mass movement of troops by road from London to Brighton. This was mainly intended as a publicity stunt by the AA, but the military were so dismissive of the exercise and motor transport in general that far from ordering hundreds of new troop carrying lorries they simply noted that if this was ever to be attempted again in the future, some sort of chin strap should be fitted to the mens service caps to stop them blowing off in the wind. Most ordinary people still travelled by horse or more likely by foot and Hornsbys were out to prove that oil engines were the future of road transport and that their engines were superior to steam. To this end in 1904 they entered one of their machines in a War Office competition to make a tractor that could pull a 25 ton (25,401kg) load for 40 miles (64 Kilometres). Hornsbys tractor actually did a further 18 miles (29 Kilometres) before the allotted amount of fuel ran out winning them the first prize of £1000. Later the same year Hornsbys Managing Director David Roberts patented a chain-track system. Roberts was similar to William Tritton in that he was not only the Managing Director of the company, he was an excellent 'hands-on' engineer and designer with a boundless imagination. In 1906 Hornsbys used Roberts track system on a range of 20 to 80hp tracked machines marketed as Hornsby-Ackroyd Oil Engines. They even tried the new system on normal road going cars, creating a tracked Mercedes and the Rochet-Schnider tracked car. The two cars were widely publicised with the Rochet-Schnider apparently being the first commercial, road going vehicle to appear on moving film. The Mercedes was capable of 20mph (32kph) and had been specifically created for use on expeditions into the Egyptian desert. In an attempt to replicate the conditions the car would face in the desert

The Hornsby produced Rochet-Schnider tracked car (The Tank Museum, Bovington)

One of the six Hornsby Petrol Tractors to be bought by the Army (The Tank Museum, Bovington)

Hornsbys undertook the new machines trials on the beach front at Skegness. Unfortunately none of Hornsbys tracked creations sold particularly well, but in 1910 Hornsbys collaborated with William Fosters in order to build the steam powered Foster-Hornsby chain-track haulage engine or the Yukon Tractor, as it became known. Unfortunately, at the time the commercial possibilities of track laying vehicles were limited and Hornsbys cannot have had a great deal of faith in an upturn in the market as the Yukon Tractor was a one-off project, no more were planned and nobody else showed any interest in ordering such a machine. Between 1904 and 1912 Hornsbys built the Hornsby Track Laying Tractor. It was similar concept to the Yukon Tractor, but on a smaller scale and equipped with a petrol engine instead of being steam powered. As before, sales of the new tractor were disappointing and they managed to sell less than ten of the petrol-engined units. The British army purchased six of them, one of which, Hornsby Number 3, still survives in running order at the Bovington Tank Museum in Dorset. After a business trip to Germany in 1911 a retired Army officer, Captain T.G. Tulloch, proposed a simple plan to create an armoured version of the Hornsby petrol engined tractor. However as is often the case with simple ideas, the plans grew until Tulloch eventually proposed a huge double ended armoured vehicle made from two Hornsby tractors which would be manned by 100 soldiers with six quick firing twelve pound guns and a dozen fixed machine guns at their disposal. Tullochs over ambitious idea never got very far and by 1911 Hornsbys experience with tracks appears to have left them unimpressed. In 1912 their patents were acquired for the relatively meagre sum of £4000 by the Holt Tractor Company of Stockton, California. Holt was in a far better position to utilise and develop track technology, having a tried and tested market for such vehicles right on their doorstep. They used the technology they had acquired from Hornsbys wisely and soon developed it to create the highly profitable

A 120hp American Holt tractor under load in France 1917 (The Tank Museum, Bovington)

The unfortunate results of simply attaching a metal box to a set of Holt tracks during American Tests in 1918

75hp four cylinder and 120hp six cylinder Holt tractors. The tractors were solid, well engineered and relatively inexpensive to buy. They were widely used by American farmers during peacetime and eventually by the allies on all fronts for towing large artillery pieces during the war and by the time of the armistice in 1918 the War Office had imported well over 1500 American Holts. The American Holt so impressed the British army that almost 500 of the 75bhp tractors were built under licence in England by Ruston and Proctor and Co LTD in Lincoln, only three of these Ruston-Holts are still known to survive today. A few days after the tanks first battle in September 1916 the Vice President of the Holt company, Murray M. Baker, told American reporters that the British army had bought 1000 Holt tractors which they intended to cover with armour plate and use in battle. Perhaps this was an innocent enough remark on his part, but the ensuing press release led to the popular assumption that the first tanks were

One of the 442 Ruston and Proctor built Caterpillar Tractors, as shown in their post war book 'Our Part in the Great War'

nothing more than adapted Holt tractors and the company was widely praised in the American press for inventing the tank. Baker had not said directly that the British tanks and the Holt tractors were one and the same, but once the story began the Holt company did nothing to suppress it, apparently they were more than happy with the publicity and increased orders they were receiving as a result. However, neither Baker nor the Holt company made any claims for recognition after the war and no mention was ever made that they should. Despite this, over 90 years later many sources still claim that the first fighting tank was actually invented in America or that the early tanks were simply armoured Holts and that without the American Holt, the British never would have invented the tank. From Murray M. Baker in 1916 right through to the present day, many internet websites and even a recent television documentary continue to perpetuate this myth. One website even states quite unashamedly that '*The tank was nothing more that a farm tractor with armor plating surrounding it. The British Mark I tank was an American built Holt tractor with steel plate attached'*. In reality, not so much as a single nut or bolt from a Holt tractor was used on any of the British tanks created during the First World War.

Whatever their collective successes and failures Holt, Fosters and Hornsbys could all agree on one thing, by mid 1915 track technology was still very much in its infancy. Despite this, track-laying vehicles were still seen by many as the only option and by this time several designs were vying for the chance to break the deadlock of the trenches. Many of these vehicles were technologically innovative, but none of them were suited to the horrendous conditions they would have to endure when they arrived in France. The new creations were vetted by the newly formed Admiralty Landships Committee; headed by the First Lord of the Admiralty, Winston Churchill. The committee viewed many weird and wonderful contraptions with evocative names such as the Pedrail Landship, the Killen-Strait tractor and even the utterly incredible, but totally useless Elephants Feet. Many of the creations the committee saw were complete failures, but they do serve to remind us of how in 1915 designers were pushing the technological and design boundaries, flailing in the dark and becoming increasingly desperate for results. However much they tried, none of these creations had yet managed to combine the two basic factors that the final machine would be required to fulfil. They would need to travel over rough, broken ground to reach a trench and then be able to cross the trench, neutralise enemy fire and continue the advance. One promising machine with which William Fosters were linked, but which never got past the wooden mock up stage was Hetheringtons Big Wheel. Major Thomas Hetherington had been involved with the story of the Tanks since his experiments with new armour plating as early as 1914. He had created a design for a machine that would have been 48 feet (14m) high, 80ft (30 m) long, over 300 tons (300,000 kg) in weight and powered by a massive 800hp Sunbeam airship engine. The scheme seems utterly fantastic and even Ernest Swinton said that the big wheel idea *'sounded like a fairy-tale'*. Despite this an official order for 6 of the incredible machines was placed and Fosters of Lincoln were contracted to create the proto-type. The vehicles 40ft (12m) wheels would have made crossing trenches, crushing barbed wire and even fording the Rhine very easy indeed. However, its unbelievable projected weight of over 300 tons and overall size meant that if it had made an appearance at the front it would have either gone down with all hands in the muddy morass or would have made such an easy target for the enemy guns that it would have been obliterated in a matter of seconds. The original Big Wheel design was eventually deemed to be unrealistic, but Tritton still thought it had potential. He and William Rigby altered Hetheringtons original scheme and came up with a smaller, lighter and hopefully more practical version to be called the 'Tritton Trench Tractor'. Instead of 40ft wheels the new version would have wheels of just 15ft (4.5m) high and instead of the original 800hp engine the Foster designed machine would be powered by a 105hp Daimler engine of the type used in Fosters heavy haulage tractors. However, as the wooden mock up took shape in Fosters yard, it became all too apparent that the new design would be far too expensive to produce and it would be just as vulnerable to enemy fire as Hetheringtons original machine. For these reasons in 1915 all orders for 'big wheel' machines were cancelled. Despite this, Tritton remained much more enthusiastic about

Tritton and Wilson inspect progress on Fosters wooden mock-up of The Tritton Trench Tractor (The Tank Museum, Bovington)

using tractors with large diameter wheels than tracks, at around the same time that he was working on the Trench Tractor he is known to have told several people that in his opinion the caterpillar system of tracks would never be any good for the job. However, just five months later he would totally redesign the track link system, making them the only viable option. For the moment, yet another promising design had been cancelled and the wooden mock up was dismantled. Fosters did manage to get past the mock up stage and create at least two excellent fully working machines prior to the creation of the tank. One of their creations could cross trenches, but could not manage rough ground and one that could sail over rough ground, but had no hope of crossing wide trenches; they were the Tritton Trench Crosser and the forerunner of the true tank, the Number 1 Lincoln Machine or Little Willie.

The Tritton Trench Crosser

During the First World War the Royal Navy were the pioneering testers of many new inventions, such as the fledgling aeroplane and the armoured car. Therefore, it naturally followed that when people started to think of new vehicles to be known as 'Landships' and 'Wheeled Dreadnoughts', the Navy was the obvious agency to be charged with their development. The Admiralty had first become involved with William Fosters after they had called for a design for a massive 15" Howitzer. The new gun was finalised and eventually created, but they had no tractor capable of pulling it. After approaching Daimler for a vehicle they were passed onto Fosters who were at that time producing the 14 ton (14,225kg) 105hp 'Silent Knight' petrol engined Daimler-Foster 'Agrimotor' tractor, mainly for the South American agricultural market. This was an excellent machine that would prove vital two years later during the creation of the tanks. After talks with William Tritton the Admiralty placed an order for 97 tractors at a cost of £1,866 each. Top brass from the Royal Navy travelled to Shoeburyness in December 1914 to observe trials of their new tractor. For some reason now lost in the mists of time, during the Shoeburyness trials the Navies latest vehicle was driven by Fosters Chief Draughtsman William Rigby. One of the dignitaries invited to see Rigby piloting the tractor was Admiral Bacon, the Commander of the Cinque Ports. He was fascinated when during the trial Rigby brought the tractor to a stop so that a team of soldiers could lay a wooden bridge over a ditch in order that it might continue. Bacon had already spoken to Winston Churchill regarding the possibility of creating a machine that could carry its own portable bridge for trench crossing. After the tests had been completed Admiral Bacon approached Tritton and Rigby to ask if it might not be feasible to adapt their tractor to carry its own bridging equipment, drop it when it reached a trench, cross over and retrieve the bridge ready for next time. This would have been just the sort of question to appeal to the two Foster representatives inquisitive minds. Just over two weeks later Tritton and Rigby had finalised their proposal, produced detailed drawings of the new machine and created a design model ready to present to the Landships Committee. Churchill and the committee were justly impressed, the design was approved and work began on the full size machine immediately. Unfortunately, just one month later in February 1915 the project was cancelled, but work on it had progressed so well that Fosters were allowed to finish the prototype. The incredible speed with which Fosters produced their new creations was to become one of the Landship Committees main

The moment of truth as the Royal Navy watches its new 105hp Daimler-Foster tractor cross a trench at Shoeburyness in 1914 (The Tank Museum, Bovington)

reasons for returning to the firm every time they wanted something new. As had already been shown with the 'Tritton Trench Tractor' Tritton had a habit of naming Fosters military creations after himself and this next machine was no exception. Presented as the Tritton Trench Crosser, modified Howitzer tractor Number 44 performed well. From the rear it looked like a standard Daimler-Foster 105hp tractor, but when the whole machine was viewed it was obvious that this was something quite different indeed. The front wheel of the original machine had been rearranged so as to run in tandem, wooden extensions had been added to either side of the chassis and the drivers position had been moved from the rear of the machine to just in front of the box radiator. A pair of thin, but sturdy, 15ft (5m) long bridges were stowed along each side and a heavyweight chain winching system had been added. Wood was used for the chassis extensions instead of steel in order to shift the tractors centre of gravity towards the heavier rear end. In conjunction with the tandem front wheels this meant that when the tractor was driven straight at a trench the front wheels would easily span the obstacle whilst still leaving the rear end on the other side. The bridges would now be deployed by hand and the rear wheels of the tractor would follow the nose over the trench. Once safely over the obstacle, the bridges would be winched back into place

Early 1915 in Fosters factory yard and the Tritton Trench Crosser goes through its military evaluation (The Tank Museum, Bovington)

again ready for the next time. The Tritton Trench Crosser was well received by the Admiralty and as with the tests at Shoeburyness, it was again driven by William Rigby. Unfortunately, the machine had two serious failings that would stop it ever going into production. Firstly, the entire bridging operation took almost 10 minutes to complete and the machine needed a clear 25 feet (8m) of working space. This was far too long to be safely stationary in the open and far too large an amount of flat, clear ground to be guaranteed under battle conditions. In addition to the machines two main limitations it was becoming obvious that the need was for a trench crossing vehicle that could not only breach trenches, but could travel over the German wire defences. The Trench Crosser would have soon become hopelessly tangled in the yards of wire on the Western Front; any future designs would have to run on tracks. Trittons Trench Crosser was an important step forward and can rightfully take its place in history as the worlds first purpose built trench-crossing machine. After its cancellation the Trench Crosser was restored to its original form as standard Howitzer tractor number 44 and returned to the Royal Navy, apparently at an extra cost of only £20.

The Number 1 Lincoln Machine

Useful lessons had been learned during the creation of the Trench Crosser, the main one being that standard wheel driven vehicles would be of little use when it came to the mud and barbed wire of Flanders. Designers, engineers and the Landship Committee alike were all in agreement that a track laying design would be the only one with any chance of breaking the stalemate. Fosters had dealt with the problem of tracks before. As previously mentioned in 1910 they had collaborated with another Lincolnshire firm, Richard Hornsby and Sons of Grantham. Between them they had produced the Yukon Tractor. This fabulous machine was basically a standard Foster 80bhp compound steam traction engine with Hornsby chain tracks instead of wheels. The massive machine was produced for the Northern Light, Power and Coal Co in the Yukon, Canada. Little is known of its working career, but in 1974 Mr. Ruddock, a Lincoln businessman and steam enthusiast was on a fishing holiday in Canada when he found the running gear quietly rusting away in a field. The machines Foster engine and boiler had been removed for use as a stationary engine on a nearby site.

The remarkable Foster - Hornsby 'Yukon" tractor.

Unfortunately since it's initial discovery in the 1970s the boiler and engine have been taken for scrap and used as backfill during the construction of a baseball diamond. Efforts to get the machines remains returned to England have failed, but the running gear of the Yukon Tractor has now been saved and fund raising efforts are ongoing to try to restore the tracks and recreate the remainder of this historic machine. Hopefully, sometime in the future the Yukon Tractor will live again and although a permanent return to England is unlikely there has been serious talk of a worldwide tour once the restoration is complete.

After their experience with the Yukon Tractor Fosters produced a petrol engined half-track machine called the Centipede. The Centipede was designed after Tritton and Rigby had visited South America in 1910. They managed to obtain an order for 46 of their small twin cylinder oil engined

A 1912 works photograph showing the Foster Centipede just prior to its despatch to South America

tractors after an impressive demonstration of the machine in which they used it to pull 21 disc shares at once. On their way back to England their boat made a stop at Brazil, where Tritton managed to take an order from a sugar plantation in Paraguay for a petrol driven chain track machine. This was quite a bold move on Trittons part, as at the time Fosters did not make any such vehicle and didn't even have any drawings. As soon as they returned to Lincoln work began on the design of the new tractor and remarkably, the Centipede was finished and dispatched in early 1913. The machine was 17ft (5.5m) long, almost 8ft (2.5m) wide and ran on a set of machine toughened steel tracks that were fully adjustable, so that they could either run flat or could be cambered to any radius, depending on the ground conditions. In 1914 an order was obtained from Java in the Dutch East Indies for fifty of these machines. However, perhaps due to the outbreak of war, this order never seems to have been completed and although the Centipede still appeared in some of Fosters postwar catalogues, the machine for Paraguay seems to be the only one of it's type ever produced. Details of the Centipedes working life and its ultimate fate are unknown, but perhaps like the Yukon Tractor, the Centipede still exists lying unloved somewhere in the undergrowth, awaiting the attentions of a very brave restorer. Fosters experience with tracked designs like the Yukon tractor and the Centipede was one of the main reasons they were chosen by the Landships Committee for the creation of the track-laying trench crossing machine in 1915. In late July 1915 William Tritton travelled to London where he was interviewed by Lieutenant Colonel Albert Stern of the Landships Committee. At the end of the meeting he sent a telegram to Fosters informing them of the results and that they should begin work immediately *'Whole Landship job handed to us with Wilson as overseer'*. William Rigby had been working on various designs for the Tank for some time and when Tritton arrived back in Lincoln work was already well underway. In addition to the new drawings a brand new set of Bullock tracks and drive gear fresh from Chicago was awaiting inspection. The Bullock Creeping Grip California Giant tractor was a similar creation to the Holt, but it differed in the placement of the engine. Unlike the Holt, the Bullock tractor had the 50 or 75hp engine placed at the rear of the machine, directly over the track unit. This placed the bulk of the machines overall weight exactly where the tank pioneers needed it, so in theory, the Bullock machine was the perfect basis for the job. Trials of experimental vehicles had been going on in Burton-on-Trent for some time under the expert eye of Walter Wilson. He had worked with the Bullock tracks before and he was not enthusiastic about the prospect of having to use them. They were splendidly designed for their intended purpose, steady use on level ground, but when they had been tested over trenches they had a tendency to sag away from the guide rollers and come off. Tritton had been characteristically

Tank pioneer Albert Stern
(The Tank Museum, Bovington)

The Bullock Creeping Grip Tractor, an excellent machine, but not a good basis for an armoured fighting vehicle

vocal about how unimpressed he was by track laying vehicles in general, so neither of them could have particularly relished using the American tracks. Despite this, of all the nations to produce tracked vehicles during the Great War, Britain was the only country not to use either the Holt or Bullock track systems in combat.

A pair of Bullock Creeping Grip Tractors on Test at Burton on Trent in the summer of 1915.

In the William Foster and Co engine log, amongst the orders for steam traction engines, there is a simple one-page entry of no more than five words recording the fact that on August 15th 1915 an order had been obtained for the construction of a Landship. This is the first time the machine that would become the forerunner of the tank is mentioned in the company records and by the time that this short note was written, work had already begun on the new vehicle. It would be little more than a riveted steel box made from standard flat boiler-plate and arranged around the chassis and engine of the existing 105hp Daimler-Foster tractor with the unaltered Bullock track unit fitted beneath it. Tritton added a wheeled tail unit to the back of the machine that was intended as an aid to steering and support of the rear of the vehicle whilst it traversed an obstacle. Secrecy was paramount to the success of the new venture and at the time Fosters were doing no official war work. Workers in factories employed on munitions and other war materials were issued with badges and certificates in order that people would know that they were not *hanging back* and shirking their countries call. As the work that Fosters were doing was absolutely top secret, they could hardly issue their employees with badges notifying everyone that they were employed on war work. Consequently, as they had no badge they could not give any plausible excuse for staying at home and many of Fosters best men became tired of being accused of cowardice and joined up. Even with the introduction of women into the factory the shortage of skilled workers to make the tanks was in danger of becoming serious, until Albert Stern took matters into his own hands. Stern saw the issuing of badges as the lesser of two evils and decided that they must be given out immediately. He would later have official complaints levied at him for his bullying tactics, but Tritton appreciated his intervention. He wrote personally to Stern saying *'I am very grateful for the trouble you have taken in this matter and I feel that I must congratulate you on the promptness with which you have overcome this entanglement of red tape'*. In reality Stern had stormed into the War Badge Departments offices in London and threatened that if the matter was not rectified immediately he would bring 20 Squadron of the Royal Naval Air Service and take them by force, needless to say Stern got his badges. The men of Fosters now had an excuse for being at home and could officially class themselves as being in reserved occupations. It was at about this time that Walter Wilson took a remarkable step forward in tank design. On August 17th 1915 he proposed that it would be better if the tracks were not on either side of the machine, such as was being used on their present design, but that they should travel all the way around the hull, resulting in the equivalent of a massive wheel on either side. A week later Tritton was in London again for an interview with the Director of Naval Construction, Tennyson D'Eyncourt, who was another Lincolnshire related link in the chain, having been born in the small town of Market Rasen, only 15 miles (24km) east of Lincoln. Tritton proposed the new idea for Willsons rhomboid design and D'Eyncourt approved it immediately. Back in Lincoln work continued on the first design in the hope that it would act as a test bed and hopefully answer some of the questions

Lincoln Number 1 takes shape in Fosters main erecting shop, note the dummy turret on the roof (The Tank Museum, Bovington)

Lincoln Number 1 machine about to go through its paces on Lincolns South Common in mid 1915. Note the covered dummy turret and the Bullock tracks (The Tank Museum, Bovington)

being raised about track technology at the time. When it was finished, the vehicle, complete with dummy turret and gun, went for trials on the South Common next to Cross O'Cliff Hill in Lincoln. Fosters latest creation was now known simply as 'The Tritton' or the 'Number 1 Lincoln Machine'. Unfortunately as Wilson had forecast the American Bullock tracks were totally unsuitable for trench crossing and spent more time being refitted than they did being evaluated. In addition to the obvious off-road failings of the original agricultural tracks, the Number 1 Lincoln Machine weighed over 16 tons (16,260kg), much more than the tracks projected maximum load. The Bullock tracks were scrapped and there ended the British involvement with imported track sets in the story of the first tanks. However, this now presented Tritton and Wilson with an apparently insurmountable problem; no track set produced in the world at that time was suitable for the ground conditions they would have to contend with or the forces that would be exerted on them. In order to work on the problem without distractions Tritton had begun to operate from the small Yarborough conference room in the White Hart Hotel on Lincolns Bailgate. Wilson still had to divide his time between Lincoln, Burton-on-Trent and London, but whenever he was in Lincoln, he and Tritton would burn the midnight oil in their temporary drawing office at the White Hart. They were determined to solve the unsolvable track problem. Tritton took the problem so seriously that he even refused to marry his fiancé, Isabel Gillies until the Landship design, including tracks, was finalised. At the end of each day he and Wilson would burn the unworkable ideas in the fireplace of the Yarborough Room, the better ideas were sent down to Fosters and tested, one after another they failed. Tritton finally had an epiphany in which he created a simple chain link with pressed steel plates riveted along its length. On the 22nd September 1915 he sent his famous telegram to the Admiralty;

New arrival by Tritton out of pressed plate **STOP**
Light in weight but very strong **STOP**
All doing well, Thank you **STOP** Proud Parents

Tritton had returned to his engineering background and gone back to basics. The Bullock, Hornsby and Holt track sets were unnecessarily complicated due mainly to their suspension and tensioning systems, whereas the new tracks were totally unsprung making them simple and hard wearing. Due to their interlocking design and basic chain tensioners Tritton also hoped that they would not sag and be thrown off when the vehicle was driven over trenches and rough ground. With the creation of the new tracks there was no reason why they could not be lengthened to travel around the entire vehicle on future

designs. The Number 1 Lincoln machine was soon given the new tracks and also acquired another name, Little Willie. It is not known for certain why the name was chosen, it is widely accepted that it was an insulting name for the Kaiser, but it is more likely to be just another of Fosters machines bearing William Trittons name, although Tritton could never have been described as little, in stature or character.

It is with the creation of Little Willie that another tank pioneer enters the story. This man was absolutely vital to Lincolns part in creating the tank although he never claimed to have invented or created anything and he is never mentioned in any of the official histories or Landship Committee reports. Few people, even military historians will have heard of him, his name was Charles Edwin Maughan and he was the worlds first ever tank driver. Maughan was a Lincoln man, living just down the road from Tritton on Lindum Terrace and working as chief tester and engineer at Fosters. In his role as chief tester he was chosen to drive Little Willie, and later the first true tank, Mother, during official trials at Lincoln and Hatfield. Charlie collected the autographs of everyone he had given rides to while driving on the tank trials with Little Willie and Mother. His book is known to have been signed by such people as William Tritton, Winston Churchill, Walter Wilson, Ernest Swinton, Albert Stern, Lord Kitchener and even King George V. The Royal family has never been known for giving out autographs and Charlie would never tell how he managed to persuade the monarch to sign his book. However, the story goes that during the Kings trip in the first tank during the 1916 trials at Hatfield Park the allotted gearsman for the day, a young Navy Lieutenant, was making such a hash of the gear selection

The driving force behind the first tanks, Charlie Maughan with one of his slightly smaller machines

that Charlie forgot himself and stopped the tank in order to lecture the poor young man on his abilities and the shocking state that the Royal Navy must be in if he was an example of one of its best young officers. Suddenly, he remembered his distinguished passenger and the Royal families ancient links with the Navy. Thinking he might be taken straight to the Tower of London, Charlie turned slowly around, but to his utter surprise the King was sitting with a broad smile across his face, highly amused at the whole situation. The ice had been broken and the King signed Charlie's book without a second thought. He would meet George V a second time when the King and Queen visited Fosters during a Royal tour of Lincolns Factories two years later. Charlie was undoubtedly a character and stayed on at Fosters after the war. He grew his own tobacco and apparently the foul smelling smoke from his homemade cigarettes could be detected above the stink of the glue factory next door. By all accounts he became a law unto himself around Fosters, calling the directors and managers by their first names and spending most of his time constructing scale model tanks from scrap metal that he salvaged from the bins. His models ran on compressed air and he would often bring the factory to a standstill by giving impromptu demonstrations of them on the foundry floor. At least two of Charlie's models still survive, a male Mark I and Tog I being on display at the Museum of Lincolnshire life in Lincoln. Charlie retired from Fosters early in 1961 and moved to Australia to live with his daughter. Right up until his death in July 1963 Charlie was still a keen promoter of the tank and the town that both he and the machine called home.

Taken outside Fosters works in early 1916. Little Willie in its final form, minus turret and running on Trittons steel plate tracks

He apparently had an uncanny talent for being able to steer any conversation around to tank production and whilst the controversy about the true inventor of the tank will probably run on forever, one thing is for certain, the worlds first tank driver was Charlie Maughan.

Both Charlie and Little Willie performed well at the trials of the new 'Tritton Track Set'. Technically the new machine was fine, but Willies design meant that he was perfect for open ground and could even cross trenches, but not trenches as wide as the military had specified or as wide as the Germans were now excavating. The machine had answered all the questions it had been built to, the controversy over the bullock track had ended and the new track design had been finalised. The way was now open for production of the next phase, the worlds first fighting tank, affectionately to become known as Mother.

Little Willie is known to have remained at Fosters until at least late 1918, but in the early 1920s he appears to have been donated to the Tank Corps and taken with several other important tanks to be stored in the Crystal Palace at Sydenham, London. When the building was destroyed in the fire of 1936 it was considered uneconomical to move the tanks to a new home and most were cut up and scrapped. Amongst the few that were deemed worth saving was Willie. The worlds first tank returned to the Tank Regiments home at Bovington Camp in Dorset where he joined other rescued machines in an embryonic museum fenced off in a quiet corner of the camp. Several stories have been circulated regarding Little Willies deeds during the Second World War, some more fanciful than others. One story goes that Willie spent 1940 defending a Gloucestershire airfield and an even more remarkable tale is that Willie was fitted with a turret from a Centurion tank and spent World War Two defending Coventry from German paratroopers. It is much more probable that Little Willie was simply parked on one of the approach roads to Bovington and used as a temporary pillbox. Willies service record for World War Two has become mere speculation, but whatever his deeds, he had survived a second war and was returned to the Tank Museum where he still remains to this day. The Lincoln Number 1 machine has now been renovated and is arguably the most important vehicle in the Bovington collection. Willie is now missing his lights, engine and gearbox, but he still survives. In a way the missing parts are of no consequence as he is a 92 year old prototype fighting machine that never saw action and the fact that he has survived at all is nothing short of remarkable.

Mother

Far from deterring Wilson and Tritton, Little Willies failings simply spurred them on to return to their work and create a tracked machine that would be able to cross rough ground, broad trenches and barbed wire. The apparently insurmountable problems regarding the tracks had now been solved and the design drawings had been finalised for the new vehicle. However, the design of the new machine would not be the only radical change that Fosters would now be facing. When Little Willie had been tested on the South Common in Lincoln, the remarkable machine had, quite naturally, attracted a sizeable crowd of puzzled onlookers and Fosters employees. Even Willie's initial journey to the improvised testing ground had not gone unnoticed by Lincolns citizens. By late 1915 Mrs. Able, the wife of the headmaster of Rosemary Lane Primary School had heeded her countries call and like thousands of other women was now doing her bit for the war effort. Mrs. Able was employed at Rustons Anchor Street timber works where she made fuselage frames for Ruston made Sopwith and Bristol aircraft. She had just finished the late shift and as she walked home over the hump-backed bridge on Dixon Street, the monstrous apparition of Fosters latest creation appeared out of the early morning mist in front of her. Although Mrs. Able would relish telling the tale of her early morning encounter in the years that followed the war, she cannot have been the only person to witness the new machines procession through Lincolns quiet streets and her story is indicative of the apparent disregard for secrecy that had amazed the military contingent of the invited audience at Little Willies initial trials on the South Common. Ernest Swinton, the militaries official 'eyewitness', was particularly vocal on the matter, although even he had to admit that by the time of the trials it was far too late to change the arrangements. Tritton would later insist that he had not been disregarding security; instead he had merely decided to pretend that the new machine was of little consequence, so in place of armed guards, Tritton had employed an open secret form of protection.

Whatever Trittons motives were, the security of the next design would be dealt with in a much more military fashion. On the 29th September William Tritton presented Fosters full size wooden mock-up of the new vehicle to the carefully selected audience gathered at the military testing grounds at Wembley. The model was hidden under a large marquee and all present declared themselves to be greatly enthusiastic regarding the potential of the machine represented by the wooden prototype. Swinton, who had been responsible for drawing up of a list of requirements that the new machine would have to fulfil had already received a sneak preview of the mock-up whilst viewing Little Willie in Lincoln a few weeks previously. He was particularly pleased and said that *'in front of me, though only in wood, was the actually embodiment of my ideas and the fulfilment of my specification'*. The day after the official unveiling at Wembley, Albert Stern told Tritton that he was to begin work on the new machine immediately. It would be lozenge or rhomboid shaped, 32.5ft (10m) long, 14ft (4m) wide and 8ft (2.5m) high. The 60ft (18.5m) long plate tracks would run right around frames attached to either side of a hull similar in shape to the main body designed for Little Willie. The chassis, engine and gearbox from the production Daimler-Foster 105hp tractor were again used for this new design. Another feature inherited from Little Willie was the steering gear that extended from the rear of the machine. The location of the engine in the middle of the hull meant that a turret would be impracticable. Therefore, sponsons extended from both sides of the track frames, after some considerable debate it was decided that these would carry standard Vickers water-cooled machine guns and 6 pound or 57mm quick firing guns of the type used primarily by the Royal Navy. The pattern of rivets used on this first machine was different to all production machines that would follow, being riveted in Boiler-makers Pitch instead of the final Girder-makers Pitch. This resulted in almost twice the number of rivets as used on a standard production machine and was done in order to give the maximum amount of structural strength possible, as at the time nobody knew what affect the recoil of the 6 Pdr guns would have on the vehicles superstructure. The final design, engine number 14225 order number AX was finished in just 36 days and had its first test run at the beginning of January 1916 at Poppletons Field and later at Burton Park, both in Lincoln. Fosters new creation was then known as 'The Wilson Machine' or 'Big

The first of the line, Mother takes a break to pose for official photographs during trials at Burton Park, Lincoln

Willie' and everyone was most anxious to find out what would happen to the machine when the 6pdr guns were fired, Thomas Hetherington and Albert Stern, decided to find out for themselves. Several rounds of solid armour piercing shot were acquired and Hetherington himself took charge of the gun and prepared to fire the first shot ever fired by a tank. This was quite a brave undertaking, given that as he climbed into the tank he apparently bet Walter Wilson £50 that the first shot would cause the entire hull to collapse. The six-pounder was loaded, Hetherington took careful aim and gently squeezed the trigger and nothing happened. He and Stern were tentatively examining the gun when it suddenly went off without warning and Lincoln Cathedral was well within range. Luckily, after two hours of searching and digging the projectile was discovered buried in the ground not far from Big Willie. More shells were fired, all without a hitch, and without any detrimental effects on the machines superstructure, it is not recorded whether Hetherington honoured his £50 bet with Wilson.

Official trials of Big Willie were held on the 13th January 1916 in Burton Park. The machine did all that the Admiralty had asked and more besides, taking trenches, boggy ground and other obstacles in its ponderous stride. On level ground the machine had a maximum speed of around 4 mph (6.5kph), on rough ground this was usually down to around 1 or 2 mph (3.5kph).

Mother during trails at Burton park, Lincoln. From right to left is William Tritton, two lady spectators, Walter Wilson, Thomas Hetherington, C.W. Pennell, William Rigby and lastly Charles Maughan about to enter the tank.

With its initial success Big Willie gained the new title of His Majesties Land Ship, Centipede, although by now everyone was beginning to call the machine Mother. Once again the problem of how to tackle security was raised. Secrecy surrounding the construction phase of the new machine could be seen as almost non-existent. Albert Stern had offered Tritton a Royal Marine guard for the factory, but Tritton had refused on the grounds that this type of high-level security would only create unnecessary suspicion and instead he would rely entirely on the discretion of his employees. His trust was well placed, and Fosters workers would be greatly praised after the war for keeping their part in the tanks development absolutely secret. In addition Tritton encouraged the popular view around the factory that he was becoming increasingly eccentric and that Mother was just one of his mad fancies that would soon be forgotten. Either way, security was now a serious consideration and the testing grounds at Burton Park and Wembley had both been deemed to be far too open to be secure, so Mother was taken for further official tests in the grounds of Hatfield House, near Welwyn Garden City, just north of London. She almost never arrived at all as Railway staff at Lincolns St. Marks Station had initially refused to let her pass through, the tracks being slightly wider than the truck she had been loaded onto. They relented after being told that this was of national importance and reminded the penalty for hindering the war effort and therefore aiding the enemy was death by firing squad.

Mother finally made it to Hatfield accompanied by her predecessor, Little Willie. On the 2nd February 1916 Mother performed in front of her toughest crowd yet, dignitaries who would watch her put through her paces included Lord Kitchener, the Minister for Munitions, the First Lord of the Admiralty and King George V. Unlike earlier tests security at Hatfield was tight with each invited audience member being required to give the code-word 'Puddleduck' before they would be admitted. Not only was security tighter, but the course was considerably more challenging than the one set out a month earlier in Burton Park. Despite this Mother, driven again by Charlie Maughan, performed faultlessly. Lloyd-George wrote *'at last we have the answer to the German machine guns and wire'*. Swinton wrote that *'she carried out the set tasks with ease'*. Lord Balfour was described as taking an almost childlike pleasure in the tank and Murray Sueter suggested that '*3000 such machines should be built immediately'*. Not everyone present that day seemed so enthusiastic about the new invention. Lord Kitchener was heard to say that the machine was *'a pretty mechanical toy, but of very limited military value, they would soon be knocked about by the enemy's artillery'*. It has been suggested that Kitchener was deliberately playing down his enthusiasm for

A useful side view of Mother, in Burton Park

Testing Mothers steering gear in Burton Park, even though the gear is hard over to the left Mother still seems to be going straight ahead (Ray Hooley)

Mother. However, even the Minister for Munitions was quoted as saying that *'He certainly gave me the impression at the time that he thought little of the invention'*. Ernest Swinton believed passionately in the new invention and was obviously quite disheartened when he said '*Kitchener had been the man I had most wished to convince'*. Just four months after the trials of Mother at Hatfield, Kitchener would be dead, drowned when HMS Hampshire struck a mine off the Orkneys on the 5th June 1916. Whatever his motives were at the Hatfield trials, Kitchener had remained rational and looked at the new invention through an experienced soldiers eyes, consequently he had immediately spotted the thing that was to become the tanks Achilles heal, its vulnerability to well placed artillery. Despite Kitcheners disappointingly logical views many of the other influential figures, including the King, were impressed with the performance and the potential of HMLS Centipede and it was soon decided that production of the new weapon would begin immediately. Secrecy was still a prime concern, especially now that the workers had their 'war service badges' and everyone knew that Fosters were employed on secret work. On Albert Sterns suggestion the term Landship was deemed to be far too descriptive and so in order to maintain security the machines were originally described as being snow ploughs for the Russian army. In an attempt to reinforce this story '*With care to Petrograd*' was written on the sides of the first tanks in Russian script, it has since been noted that this was incorrectly spelled. A more believable cover story was that the new machines were mobile water carriers designed for the rough country of Russia and Mesopotamia. Aware of the military penchant for describing things by their initials Albert Stern wanted to eradicate the term water carriers as he did not relish becoming the secretary to the W. C. Committee, W.C. also being short for Water Closet, another name for the Toilet. He claims to have changed the title from water carriers to water tanks instead, this was shortened to a single word and the tank was born. However, in his short article '*The evolution of the Tank'* William Rigby tells a very different story, saying that '*due to frequent modifications to Mothers hull the water carrier soon became known around the foundry as that bloody water tank, soon to be simplified to tank'*. According to Ernest Swinton if the Germans had been the first to build a viable machine the term tank would never have existed and we would all be calling them '*Schutzengrabenvernictungpanzerwagen*'.

After her successful trials at Hatfield, production of the machines made in her image began and Mother was returned to her home at Lincoln. However, Mothers war was not over yet and when a machine was needed at the newly formed tank training ground at Bernesfield, it was Mother that was despatched. At Bernesfield she started to be treated as a lady of her standing might reasonably expect and was repainted in a glorious impressionist 'landscape' by the famous Royal Academy artist Solomon J. Solomon. The design was Solomans idea of camouflage, the scheme was intended to break up the tanks silhouette and Mother was used as an artists model to teach soldiers how to paint their own machines when they arrived in France. However, after a few days at the front the tanks would have been naturally camouflaged with mud and the Solomon design was soon dropped, only appearing on a few of the first Mk I machines. After her brief spell as a model, Mother soon began to be treated quite differently. During these early years the supply of ordnance was at best slow, but more usually non existent and so Mother had to surrender her 6 Pdr guns for fitment into a fighting tank. Worse was yet to come for Mother when in March 1917 the Oldbury Trials were held in which companies who thought they could come up with a better tank transmission system were each allocated an old machine on which to experiment. Mother was issued to the Daimler company who modified her to incorporate their Petrol-Electric drive system. Although a similar system is used on many tracked vehicles today, at the time the Daimler proposed unit was beset with teething problems and after the trials little is heard of Mother. She must have been stored somewhere as she next appears in about 1920 at Bovington Camp as part of the embryonic outdoor museum collection. Mother had been ill treated and by the time she reappears she is little more than a shell. She has been robbed of all her useful parts and has no tracks, sponsons or a roof over her hull. Clearly nobody considered her worthy of being saved and sometime before 1940 her story comes to an inglorious end, what was left of Mother, the worlds first tank, went under the scrapmans torch, or did it? As with many of histories great disappearances, rumours persist that she is not lost at all and that Mother survived the torch. The story goes that she was not scrapped at all and in an attempt to hide her away she was buried somewhere on Bovington Camp, along with one or two other priceless prototype tanks that were no longer needed, but nobody knows for sure.

No way to treat a lady, the last known sad photograph of Mother at Bovington Camp just before she disappeared (The Tank Museum, Bovington)

CHAPTER II

Tank Production

Little Willie and Mother had done all that had been asked of them and now it was assumed that serious tank production could begin. However, in his book *Eyewitness,* Ernest Swinton summed the situation up well when he said that *'after the successful trials Mother was sent back to Lincoln, there then ensued a lull while G.H.Q., whose motto apparently was softlee, slowlee, catchee monkey, was making up it's mind whether it wanted tanks or not'.* On Christmas Eve 1915 Albert Stern presented a paper to the Committee of Imperial Defence entitled *'The Present and Future Situation Regarding the Provision of Caterpillar Machine Gun Destroyers or Land Cruisers'.* In the paper Stern stated that *'to whatever department the production of the Caterpillars is finally allotted, it is suggested that it be entrusted to one businessman of proved capacity, preferably one who has been connected with the experimental construction up to date'.* This inclusion was clearly intended to warn against the instigation of yet another committee and to place the entire production in the hands of William Tritton. In the exact opposite the Tank Supply Committee was formed on the 12th February 1916 to oversee the production of the new weapon. On the day that the new committee was formed they placed an order for 100 machines to be delivered as soon as possible. Mother had been created only a few months earlier in Lincoln by William Foster and Co Ltd. However, Fosters were a relatively small concern and had only enough available space to produce 23 tanks, the remaining 77 machines would be built from Rigbys drawings by the much larger Metropolitan Carriage, Wagon and Finance Company of Birmingham. Apparently, Tritton and Rigby protested greatly at the fact that all their hard work would now be shared by other companies, but in the end it came down to a simple question of logistics and the drawings were sent to Birmingham. The original order of 100 tanks was soon increased by a further 50 machines and although Fosters were still at maximum capacity with their normal traction engine production they still managed to add another 13 tanks to the 23 machines from the original order. Fosters produced a total of 36 Mk I machines from July to October 1916 bearing serial numbers 739 to 775 on their hulls and using engine numbers 14234 to 14270 inclusively. The remaining 37 Mk Is were added to the 77 already being built by Metropolitan. According to Ernest Swinton the serial numbers began so high in an attempt to mislead. If they had started at 1 or 100 everyone would have known this was a new machine, whereas if over 700 had already apparently been built and nobody had noticed, then the thing must be of little consequence. It was at about this time that it was decided that half of the 150 tanks ordered would carry a pair of 6 Pdr guns and three machine guns, as Mother had. The other half of the order would be fitted with different sponsons redesigned by William Rigby to carry machine guns only, the new variant having a collective armoury of a pair of water cooled Vickers machine guns on either side and one in the usual place at the front between the driver and the commanders seats. This adaptation made the redesigned machines just over 1 ton (1020kg) lighter than their more heavily armed colleagues. In order to distinguish between the two types they were originally to be called 'Destroyers' and 'Consorts', but instead they were titled male and female, the more intimidating males carrying the 6 Pdr guns and the females being equipped with machine guns. One would assume by her name that

A captured female Mk II and its proud new owner, taken after the battle of Bullecourt in March 1917

Mother, was a female, but under this new ruling the tank would be classed as a Male. Swinton suggested that Adam would be a more appropriate name for the first tank, now that it was being classed as a male, but the name remained unchanged and Mother continued to be the only female tank ever to be built with twin 6 Pdr guns. In addition to trying to change Mothers name Swinton claims to have coined the term Male and Female for use with tanks, however the derivation has apparently been used on Lincolnshire farm carts for almost 200 years. There was even a hermaphrodite cart that could be switched from one to another. Therefore, it is perhaps more likely that the term was first used by someone professionally involved with both Lincolnshire farm machinery and the invention of tank, probably someone from Fosters, or to be precise, William Tritton himself.

All of the heavy tanks produced by Fosters throughout the war were exclusively male, as the factory had neither the space nor the work force needed to create two different variants, although the company did produce the majority of the designs and engineering drawings from which the major machine modifications were produced. The road to production had not been easy for Wilson, Tritton and the other tank pioneers, but now that the thing was made and production had started it was assumed that life would become easier. However, the tank would have to fight for its survival before it even began to fight in France. Many of the military top brass, lead mainly by the Cavalry, felt that the creation of such a weapon was quite unnecessary, horses had been used successfully for hundreds of years and why should things change now. In addition to the Military's objections, even those supposed to be supporting the tanks seemed to be obstructing production. Committees had been set up in order to oversee every aspect of the war and infighting between them or their members often led to childish deadlock between departments, suppliers and manufacturers.

One of the most pressing problems facing tank production was the shortage of men eligible to fight at the front. The wearing of a war work badge showed that a man was in a reserved occupation, but by 1916 the recruitment centres were empty and threats were made that skilled men, working in factories such as Fosters, might have their badges taken away in order to free them for active service. It would be a long time before tank production would be regarded with the same importance as artillery, munitions and aeroplane production was. In late 1916 Field Marshall Haig held a dinner at which Albert Stern, Tennyson D'Eyncourt and William Tritton were present. After the brandy and cigars, Haig congratulated the tank pioneers and told them to '*go home and build as many tanks as you can'*. Unfortunately he ended by adding the proviso '*subject to not interfering with the normal output of aircraft, railway trucks and locomotives*'. This apparent disregard for the pressing need for tanks was perhaps best demonstrated when in March 1917 William Tritton asked the Ministry of Munitions if he could be immediately granted an extra 250 skilled steel workers and fitters to bolster the constant influx of local and itinerant unskilled workers, eventually he was granted just one extra man. Tank Corps soldiers were often sent on special detachment to work at William Fosters and Co Ltd in order to get first hand knowledge of the tanks and their production from rough castings to finished machines. Unfortunately, one extra fitter and the occasional visiting soldier was not enough to keep up with the orders and the permanent work force were asked to work longer and longer hours until it was not uncommon for them to be doing 12 or 14 hour days. This all had the effect of demoralising the workers leading to a definite drop in production quality. In turn this lead to mechanical reliability problems in the field, a factor that was greatly appreciated by those who opposed the new machines. However, perhaps the most serious threat to the tank was that in the early years the British Army simply did not seem interested in the slightest. In October 1915 Lord Kitchener was approached concerning the creation of armoured fighting vehicles and was described by Colonel Maurice Hankey of the Committee for Imperial Defence as being *'not receptive to the idea and very preoccupied'*. Colonel Ernest Swinton was as bemused as anyone when in 1915 he commented to Albert Stern *'This is the most extraordinary thing I have ever seen. The Director of Naval Construction appears to be making Land Battleships for the Army, who have never asked for them and are doing nothing to help'*. The tank and eventually the Tank Corps would continue to have to fight on two fronts, at home and abroad until the end of hostilities. Every failure of the new machine was seized upon by those who opposed it and after the war there were calls for the Tank Corps to be disbanded saying that tanks were not wanted in the first place and now that peace had been declared they were a redundant branch of the fighting services. However, no matter how hard they plotted its downfall there were still others who saw the tank as the right tool for the job. Stern wrote that the tanks '*look like a joke by Heath Robinson, one forgets that these things have already saved the lives of many hundreds of our soldiers*' During the initial months of production in 1916, even before the tank had seen action, consideration was being given to designs for the next generation of vehicles. In July 1916 Tritton came up with drawings for an unusual machine known as the Flying Elephant. Originally named the Fosters Battle-Tank it was to have been a monster weighing over 100 tons (10,200kg), fitted with six machine guns, a single forward facing gun and state of the art 2inch (50mm) thick armour plating. This is perhaps not too impressive by modern standards, but it must be remembered that the standard production tank of the time had a maximum armour thickness of just half an inch (12mm). The Flying Elephant was designed as a response to growing fears that the Germans would soon be developing tanks of their own and was to have been the worlds first purpose built 'tank killer' boldly described by Albert Stern as '*a tank capable of resisting field guns*'. The front facing gun of the Flying Elephant is usually described as a 6 Pdr of the type fitted to the Mark I, but it would seem odd that a dedicated tank killing machine was intended to carry

A design model of Trittons Flying Elephant, full of good ideas, but the finished machine was never put into production (The Tank Museum, Bovington)

half the armourment of a standard machine? It is much more likely that the elephants main armourment was to have been something larger, probably a 12 Pdr cannon. Even when the weight was paired down to a trim 60 tons (61,000kg) it was obvious that a very special engine would be needed if the thing was ever going to move under it's own power. Eventually, Tritton settled on a hybrid version of the old 105hp Daimler engine that would have basically been two standard engines bolted together running from a common crank. The massive tank would also have carried a double set of tracks underneath the main body that would have supported it on rough ground. This was a feature that Tritton patented as the intriguingly named 'Live Belly' system. In the event neither the Flying Elephant nor Trittons Live Belly would ever see action as the design was cancelled by the committee and the half-finished prototype was scrapped.

Towards the end of April 1916 Sir Douglas Haig sent a communique to the Tank Supply Committee requesting as many tanks as possible to be ready by 1st June of that year. This was the first indication of the tanks debut on the battlefield of the Somme. However, Haig was asking the impossible as the tanks would need fully trained crews and recruiting had hardly begun let alone training. Eight men were required for each tank, a commander, a driver, one gearsman on each side and four men to load and fire the guns. Curious advertisements started to appear in the motoring press of the time asking for men who were '*knowing with engines*', but even if sufficient men could be enlisted to crew the tanks, guns were the next problem as even these were in very short supply. Even the ammunition for the six pound guns was a scarce commodity until someone from Armstrongs remembered that they had recently supplied 25,000 armour-piercing six pound shells to the Japanese Navy, negotiations began immediately and the shells were soon on the next boat back to England. A message was returned to Haig saying that the best they would be able to manage by June would be just a handful of fully crewed tanks and only then if sufficient ordnance and ammunition could by found for them. Haig could not argue with the figures and was forced to abandon his plans to use tanks that Summer, instead he pushed for the machines to be ready for the renewed Autumn offensive.

The Tank Corps had to make do with plywood and canvas when the real thing wasn't available. This dummy Mark I Female could have been used either for training or to confuse the Germans

The history books say that the first tank engagement was on the 15th September 1916, but in actual fact the tanks went into battle almost three months before, in late June 1916. Admittedly, there were no Germans present and this battle was at the tank training camp at Elveden with the tanks fighting against each other in a full dress rehearsal for the real battle to come. The 'Battle of Elveden' was watched by a crowd of interested parties and invited dignitaries, afterwards they were all more than enthusiastic about the possibilities of tank warfare. Lloyd George remarked to William Tritton that *'if the tanks proved as great a success in actual warfare as their performance today has lead me to expect, then their influence on the war will be beyond present conception'.* The dress rehearsal was over and on the 15th September 1916 the tank entered its historic first conflict at Flers-Courcelette on the Somme, this was not a dress rehearsal, this was the real thing and unfortunately Lloyd George and many others were to be disappointed. The tanks did not perform as well as had been hoped and the poor ground conditions were hardly perfect for the new weapon. Perhaps if Haig had been able to use them for his Summer offensive as he had planned, the Tank Corps regimental history may have read very differently. A total of just 49 tanks could be mustered for the battle, 18 of which had been built in Lincoln. In order to make individual machines discernable in battle many of the first tanks were given names beginning with their Company letter. Tanks from C Company bore the names of various drinks, such as Champagne, Créme de Menthe, Chablis and Cognac. Whereas many of the tanks in D Company received more warlike names such as Die Hard and Daredevil. This tradition of naming tanks according to their battalion letters would continue and eventually female machines would receive girls names, such as Artful Alice, Deborah, and Flirt, whilst the males usually received more masculine or fearsome titles such as Black Prince, Exterminator and Havoc. Despite the carnage and swift victory that these evocative titles might have promised, the new weapons technical failings and hurried construction would soon become all too evident to those who were set to crew them, of the 49 tanks massed for the Battle of Flers, 17 broke down or got ditched while still behind British lines leaving only 32 that made it as far as their starting points. One of the 17 that would have to wait and fight another day was Foster built D7 carrying

Lincoln made Mk I, production number 742, arrives brand new at the Tank Corps training ground at Bernesfield in mid 1916. The Russian script on the side translates as 'With care to Petrograd' and was meant to enforce the story that the tanks were water carriers bound for eastern Europe (The Tank Museum, Bovington)

Tank 742 now in camouflage and bearing the battalion number D7 photographed some time after it had been ditched at Flers.

the production number 742 and commanded by Lieutenant A.J. Enoch, late of the Notts and Derby Regiment, who would eventually survive the war finishing as Captain A.J. Enoch M.C. For those tanks that did make it as far as their starting points the ground conditions were even worse than had been expected and once the battle commenced only 18 played a real part in the fighting and, as Kitchener had predicted, losses to enemy artillery were high with 10 machines soon being put out of action by German field guns. After the tanks first encounter an aeroplane of the Royal Flying Corps flew over the battlefield to report back on the days outcome. The Observer, Lieutenant Phillips of No.3 Squadron, returned to the British lines and dropped a message saying '*tank seen in the main street of Flers followed by large number of our troops'*. By the time this news reached the British press, it had been embellished and Lt. Phillips was reported as saying that '*a tank was walking down Flers high street with the British army cheering behind it'*. This report was flashed around the world, sealing the tanks celebrity status and beginning a fascination between it and the public that would endure till the present day. The tank that started it all by walking into Flers that day was D17 'Dinnaken' commanded by Lieutenant Hastie of the Highland Light Infantry. Dinnaken bore the production number 759 showing that this famous tank was a William Foster Mk I. Lt. Hastie and his crew had done good work that day in their Lincoln built machine accounting for many of the enemy including machine gun positions and snipers ensconced in the rubble of the Village of Flers. However, as Hastie and his driver Cpl Sheldon manoeuvred

The end of Dinnaken, the tank that walked down Flers High Street is unceremoniously broken up by German prisoners (The Tank Museum, Bovington)

the tank through the debris strewn streets and out of the village, the stress on the Daimler engine started to tell and the engine began to knock badly. Hastie realised that he might not have long before the engine broke down completely and quickly decided to get his tank off the road and into a place of relative safety. He managed to stop the machine safely and evacuate his crew just outside Flers. Hasties consideration for the safety of his crew and his well balanced approach to warfare would serve him well, as he survived the Battle of Cambrai a year later and would eventually survive the entire war, finishing as an acting Major and holder of the Military Cross. His unfortunate tank, Dinnaken, remained where he had left it for some considerable time, but continued to serve as a pillbox and headquarters for various officers who were commanding battalions in the front line north of Flers. Eventually though, Dinnaken, the Lincoln made machine that had first ignited the publics fascination for the tank, was broken up by a work party of German prisoners of war. Despite heroic actions such as that of Lt. Hastie and his crew Flers-Courcelette would be the first in a string of occasions that the tank would be misused in small numbers and in horrendous conditions. It would not be until fourteen months later in November 1917 that the fledgling Tank Corps would get a chance to show their full potential when hundreds of tanks were employed on firm unbroken ground during the Battle of Cambrai. However, after the hard fought battle at Flers, Cambrai was a victory that the tank crews could only dream of, but perhaps even worse than the poor conditions encountered at Flers was the fact the tanks initial element of

Lincoln made Mk I C.15 ready to go into battle for the first time at Flers. Note the anti-grenade netting over the top of the machine.

surprise had now been lost. Swinton had written as early as March 1916 warning of wasting the tanks potential in its first encounter saying that *'since the chance of success of an attack by tank lies almost entirely in its novelty and in the element of surprise, it is obvious that no repetition of it will have the same opportunity of succeeding as the first unexpected effort'*. The day after the tanks disappointing showing at the Battle of Flers Churchill wrote *'my poor land battleships have been let off prematurely and on a petty scale'*. The brilliant designer and engineer Harry Ricardo said that *'the whole operation was a tragic example of mismanagement, muddle and misunderstanding. A disaster of the first magnitude'*. These inglorious beginnings were a disappointment to the tanks supporters and its creators, but an absolute gift to those who opposed the new weapon. However, its impact on the men who watched the tank lumber into battle that first day was quite different. There are stories that when the tank went into battle for the first time the secrecy surrounding them had been so thorough that fighting stopped as the infantry of both sides stood with their heads above their parapets and simply stared at the machine having no idea what to make of it. It is also well documented that many Germans, quite understandably, surrendered without a fight when confronted with the fearsome new weapon. Although many enemy soldiers did surrender, numbers of prisoners taken during the fighting

Fascination overtakes fear as German soldiers begin to learn the tanks strengths and weaknesses

Comical, but not far from the truth as for the time being, the remarkable new invention had the Germans on the run

have been much exaggerated over the years and it is often said that thousands of Germans simply panicked and ran, after one of them saw the tanks approaching and frantically passed the word along the trench that *'the Devil was coming'*. The fact and fiction of the tanks first engagement have become inextricably entwined since 1916, muddying the truth forever. What is sure about the tanks first showing is that the fascination for the new weapon soon swept through the Army and across the channel to the home front. Ridiculous Stories began to circulate about the futuristic new machine, each one carried a crew of 400 men and could travel over any terrain at 30mph, they could spin around so fast that they were able to dig themselves into the ground, the new weapon was actually constructed in Japan by Swedish workers, but most annoying of all for the tank crews was the tale that the new machines were operated by airmen who had lost their nerve and could no longer fly. True or not the publicity surrounding the tank was unceasing and it was not long before they began to be billed as the war winning weapon everyone had been waiting for.

A Mk I female being broken up by German Prisoners, note the Russian Script on the side

Following the battle of Flers-Courcelette Douglas Haig was full of praise for the Tanks, recognising that the disappointing outcome would have been very different if the conditions had not been so poor and greater numbers of machines had been used. He said '*we have had the greatest victory since the Battle of the Marne this is due to the tanks. Wherever the tanks advanced, we took our objectives and where they did not advance we failed to take our objectives'*. Haig had such faith in the new weapon that he placed an immediate order for a further 1000 machines. Meanwhile back in England, tank production continued, but it was still a very long way from 1000 tanks. This was a busy period for William Tritton as not only was he sitting day and night by his office telephone waiting for news from Flers, but just a week later on the 21st September 1916 he kept his promise and married Isabel Gillies at St Gabriels Church, Willesden, the marriage was witnessed by Fosters Chairman C.W. Pennell. Tritton had no time for a honeymoon and the very next day he was in France inspecting the tanks under battlefield conditions prior to their attack on Theipval on the 25th and 26th September. Upon his return to Fosters it was straight back to work and no sooner had he returned to his desk than he was appointed as Director of Construction of the Mechanical Warfare Supply Department. Despite this elevation, Haigs huge

One of Fosters brand new Mk I tanks being evaluated at Robeys (Ray Hooley)

A rear view of showing both of the Robey evaluation tanks. Note the cables and hydraulic jack that will operate the tail unit when it is fitted (Ray Hooley)

order for new tanks was still quite impossible. After the Oldbury trials it was suggested by Albert Stern that Fosters should be requisitioned under the Defence of the Realm act, this would have effectively left Tritton and his team free to continue research and production backed by everything the Ministry had to offer, in the event Fosters became a controlled establishment. This change in status lead to the factory being eligible for government allowances and by the middle of 1918 Fosters had spent a remarkable £58,000 buying land for a new testing ground and buildings for expanded production. They also spent £12,000 on new tools and machinery alone. William Fosters expenditure on tank related projects related to well over half their total capital. Metropolitan also had the capacity to expand production, but if 1000 tanks were to be completed other manufacturers would have to be found. Four more firms were enlisted to produce tanks from Fosters drawings, they were Armstrong-Whitworth in Newcastle-on-Tyne, the Coventry Ordnance Works, Beardmores and Mirrless Watson and Co of Glasgow. Seven Mk1 Tanks were ordered from Robey and Co LTD on Canwick Road in Lincoln in March 1916. A brace of new Mark I tanks and a huge bundle of drawings were sent from Fosters to Robeys for evaluation, but in the end it was decided that they had neither the space nor the spare resources for even small scale tank production as they were already overstretched keeping up with the demand for aeroplanes for the Royal Flying Corps, one in fourteen of all the aircraft used during the Great War being produced in Lincolns factories. Meanwhile, production of the Mk 1 tank continued at the six factories now involved in its creation. Haigs order for 1000 tanks was deemed to be unrealistic due to material and ordnance shortages and was cut by the War Office, but by now work had begun on an improved design. An order for 100 of the new tanks, to be designated as the Mk II, had been received and was to be shared between Fosters and Metropolitan, Fosters making 25 male tanks and Metropolitan producing 75 females. This was to be followed by an order for 50 male and female Mk III tanks, all to be built by Metropolitan. The 25 Mk II tanks made by Fosters carried the serial numbers 776 to 800 with engine numbers running from 14272A to 14296A and were built from January 4th to the 13th March 1917.

In January 1917 eight Mk I machines were put on the boat from England to help the Egyptian Expeditionary Force with their forthcoming attack on the Turks in Palestine. By this time the steering gear was known to be near to useless and had been removed from all of the Gaza Mk Is prior to their leaving Bovington. The tanks also appear to have been repainted in a light yellow sand colour although the colour difference could well be due

Six of the eight Tanks operated by Major Nutt in Palestine

to bleaching of the khaki brown paint and stronger light overexposing the photographs. Whatever the reason for the apparent colour difference it is odd that the tanks used in the desert do not appear to carry any works production numbers or battalion markings from their previous lives. Therefore there is no way of knowing if the likes of Sir Archibald, Pincher or Tiger were Lincoln made machines. Likewise we will never know if they were training machines or if they had already had a glittering wartime record in France. The eight desert tanks were under the command of Major Norman Nutt from E battalion. During March and April the tanks were in a bitter struggle with the Turks near Gaza. Unlike the Germans, the Turkish army did not seem particularly worried about the tanks when they first attacked. Perhaps they had already received instructions from the their allies on how to deal with the new machines as they simply turned their artillery on the advancing forces leaving several machines badly damaged. As before, the tanks were misused with the established local British officers not willing to admit that the young Tank Corps officers knew more about tank warfare than they did. Instead of a concerted attack on a single stretch of the Turkish front line the tanks were separated into pairs and spread over far too wide an area. By mid 1917 Major Nutt's Gaza detachment had enjoyed only limited success and were left with just five tanks out of the original eight. The number was brought back up to fighting strength with the arrival of three new Mk IV tanks in late 1917. When the rag tag assortment of machines went into the Third Battle of Gaza in November 1917 they eventually showed their worth with the battle being ultimately successful despite the usual blunders from the non Tank Corps officers. The worst thing about operating tanks in desert conditions, then as now, is the sand. It blows into every nook and cranny mixing with grease and

Three Mk I machines being overhauled prior to the Third Battle of Gaza

oil to form a fine grinding paste. To combat this Major Nutt decided to run the tanks without grease. This stopped the sand from forming grinding paste, but did nothing for the already worn out track rollers. Constant maintenance and limited battlefield success combined to seal the fate of the Gaza tanks and after the third battle they were never used again. There are several stories that they survived and were left sitting in the dunes after they had outlived their usefulness. This is unlikely as any surviving Great War vehicles would surely have been found by now and probably have made their way to the armoured vehicle museum at Latrun.

The Mk 1 had been badly used in two separate theatres of the war, but by now a new design was coming off the production lines. The new Mk II and Mk III machines looked basically the same as the original Mk I, with one main visual exception; as with the Gaza Mk Is their tails had been docked. Trittons steering gear for the first tanks had been a splendid idea in theory, but proved to be a liability in battle. Far from aiding steering it got caught up in the barbed wire, slowing the machine down and making steering even more difficult. The steering gear came in handy for carrying stores, but otherwise it was useless and so only appeared on Little Willie, Mother and the Mk I machines. The new marks also differed greatly in a number of less obvious, but very significant ways. They had been created to test several proposed improvements such as a larger escape hatch in the roof, wider tracks and a revised interior.

In addition to exterior design changes the reliable old water-cooled Vickers machine guns were swapped for the American designed Lewis Machine Gun. Thousands of the new guns were built under licence in the U.K. by companies such as Rustons of Lincoln, who in addition to their normal and war production had built around 8,000 Lewis Guns by 1918. The Lewis gun was a cheaper and much more plentiful weapon than the Vickers and was popular with infantry, pilots and machine gun crews alike. It was air cooled, easy to strip and clean, whilst being light enough to carry and easy to operate. However, the peculiarities of tank warfare were only just beginning to be understood and weaponry was one of the most difficult factors to balance effectively. For instance, normally only officers would be issued with revolvers, the other ranks would carry the tried and tested Enfield Number 3 Mk I bolt action .303 rifle. However, the cramped interior of a tank was no place for seven men carrying rifles and fixed bayonets. Therefore, common sense prevailed and all ranks in the Tank Corps were treated equally and issued with the .455 calibre Webley service revolver. The Webley was undoubtedly the perfect gun for the job, unfortunately, the new Lewis machine gun was not and was quite unsuitable for work in tanks. The main problem with the

Lewis was that machine gun parties on foot were usually hidden from view and damage to their guns from enemy fire was rare, but tanks were intended to get as close as possible to the enemy and concentrated hostile fire could quickly damage the Lewis Guns jacket and barrel. It was not unknown for a female tanks entire weaponry to be put out of action by damage to the Lewis gun jackets. In late 1917 the Tank Corps even went so far as to issue an ultimatum to the Tank Supply Committee saying that the Lewis guns barrel size was unsuitable for use through loopholes, especially the one at the front between the driver and the commander and unless alterations were made no further tank actions would be fought, the entire Tank Corps would effectively be on strike! The problem with the front loophole was not remedied until the introduction of the Mk V machine, but despite the problems, the old Lewis gun remained. However unsuitable for the job the otherwise excellent Lewis gun may have been, it was a fact of life that ordnance was in short supply, so for now, the tanks would have do the best with what they were given. The new guns were just one in a long line of alterations and redesigns for the new Mk II and III tanks, but perhaps the most striking feature about the new marks of tank was that they would be intended purely as training machines. As training machines the MkII and III tanks would not be required to come under enemy fire and were consequently produced using untreated standard boiler-plate with the result that the finished machines were unarmoured. This would have been fine if the Mk II and Mk III tanks had been restricted to crew training only, but the chronic shortage of complete Mk I fighting machines for the start of the spring offensives of 1917 meant that several unarmoured and by now worn out Mk II and III training tanks were fielded in order to make up the numbers for the battle of Bullecourt in March 1917. The hulls

After the battle, Foster made Mk II male 799 is captured by the Germans

Tank 799 being raided for souvenirs after being captured during the Battle of Bullecourt in 1917

of the training machines could easily be penetrated by concentrated standard small arms fire and after the battle a pair of the new tanks, one male and one female, were salvaged by the German army. One of two captured machines was tank number 799 built in Lincoln by William Foster and Co Ltd in late February 1917. Both of the captured tanks were widely photographed on the battlefield by German photographers. Oddly the identification number on 799 seems to only appear on the starboard side and the Germans usually chose to photograph it from the port side. At the time of 799s capture the Germans were unimpressed with the new weapon and thought that tanks were of very limited military value. When they had finished their evaluations, they destroyed 799 by blowing it to pieces. Recently several large fragments of 799 were ploughed up by a French farmer, the remains of this Lincoln made tank were saved and can now be seen by arrangement at Jean Letailles private museum in Bullecourt.

It is never advisable to let the enemy in on your secrets too soon and at the time the capture of a pair of tanks of a new variant was looked on as a major breach of security. However, in this case it turned out to be quite fortuitous. The Germans examined the battered machines, taking special note of the location of the fuel tanks and the considerable amount of new ventilation holes their troops had inflicted on the hulls of 799 and it's unfortunate comrade. They never realised that the captured tanks were made from unarmoured steel, and concluded that their standard machine gun ammunition was sufficient to stop the fearsome new British weapon and that troops should aim specifically for the vulnerable fuel tanks, thus with any luck causing the machine to explode. Unfortunately for the Germans, this information coincided almost exactly with the introduction of the Mk IV. The new tank was released in May 1917 and was more heavily armoured than the Mk I. The Mk IV incorporated all of the new features tested on the MK II and III, plus a few more besides. As the Germans had realised from the study of the two machines captured at Bullecourt, the previous designs of tanks had a gravity fed fuel supply from reservoirs carried at the front of the track frames higher than the carburettors. This arrangement was fine unless someone was shooting at you with a field gun and after the capture of the two tanks the enemy now knew exactly where to aim. However, the Mk IV machine had a completely redesigned fuel system with an electric fuel pump and a larger armoured petrol tank located on the outside, hidden from view at the rear of the hull. The sponsons were redesigned to give better ground clearance and by making them retractable the tanks were easier to move by rail. A Mk I tank with sponsons was over 14 feet (4 metres) wide, much wider than any standard gauge railway, consequently, on the first machines the sponsons had to be removed prior to travel and replaced at the destination. This was no mean feat as they would count for a considerable part of the machines overall weight and after prolonged use the tanks hull would become twisted making it almost impossible to line up the bolt holes between the sponsons and the

A new female Mk IV, a vast improvement on the earlier tanks

body. Many tanks ended up with a series of new holes, drilled in order to compensate for the warping of the external plates. The guns of the new tanks were also changed again. In spite of protests from the Tank Corps the Lewis Gun remained as a replacement for the Vickers machine guns, but the 6pdr cannon on the male tanks had its barrel length shortened. This alteration had little effect on the muzzle velocity of the weapon and was primarily done

A good view showing Mk IV tanks in various stages of construction at Fosters works in late 1917

Male Mk IVs leave Fosters by the train load in late 1917 (The Tank Museum, Bovington)

in order to stop them digging into the ground or becoming entangled in the barbed wire as the old full length guns had a tendency to do. The shorter barrel was also easier to retract into the sponson, another aid to railway travel. The redesign of the sponsons also gave a chance to add another much needed safety feature to the new machines. Larger doors were added to the rear of the male sponsons and on the sides of the females. This did not stop crews becoming trapped inside burning tanks, but it did at least give them half a fighting chance of escape. When the machines were not in battle the redesigned doors could be left open to allow some of the considerable heat and fumes generated inside a tank to escape. Fosters built a total of 101 of the new Mk IVs using engine numbers 14297M to 14397M inclusively. The Mk IV tanks produced at Fosters carried a four-figure identification number on the side of their hulls running from 2300 to 2400. At least a quarter of Fosters entire Mk IV production run saw action at the Battle of Cambrai in November 1917.

Canadian soldiers hitch a lift from Lincoln made Mk IV 2365

Tank number 2300, the first Mk IV from the Foster production line was despatched under it's own power to Clayton and Shuttleworths for evaluation. Claytons were another established Lincoln firm, but were on the opposite side of town to Fosters and the tank apparently took the best part of a day to get there. Unfortunately, the great distance across cobbled streets and tram lines appears to have taken it's toll on the machines tracks and the port side track broke on the Spa Crossing just a matter of yards from Clayton and Shuttleworths front door. A lorry carrying a team of men and tools was despatched from Fosters to fix the broken track, which took several hours to complete. In the event Claytons were far too busy producing Sopwith fighters and Handly Page bombers amongst other things and could not feasibly take on the construction of complete tanks. Instead they constructed a brand new, purpose built drop stamping shop especially to make parts for tanks, although it is not known exactly what the parts were it is thought to have been track plates and linkages. Claytons may have been far too busy to produce tanks, but their aircraft were beginning to make a real difference on the Western Front. Roy Brown, the pilot officially credited at the time with shooting down Manfred Von Richthofen, better known as the Red Baron, was flying a Lincoln made Clayton and Shuttleworth Sopwith Camel number B7270 and apparently sections of the Barons downed Fokker were actually brought back to Lincoln and displayed in Clayton and Shuttleworths factory foyer.

Fresh from the production line, Tank 2308 is shown here on the testing ground in Lincoln at the William Foster and Co Ltd open day in mid 1917 (Ray Hooley)

Tank 2308 goes through its paces for the crowds, but by November 1917 it would have joined the Tank Corps and be G. 2 'Grouse' of G Battalion. The tank eventually came to an inglorious end, bellied on a tree stump in Haveringcourt Wood during the Battle of Cambrai (Bill Baker)

Note the skull and cross-bones on the top of 2308. This can be seen on photographs of Fosters Munitionettes taken on the same day, see Appendix I (Bill Baker)

By the time that the Mk IVs were being introduced it was becoming apparent that the female tanks were much more useful in combat than the males were. There was little else to shoot at on the battlefield but enemy infantry and for this reason the Mk IV started to be built on the ratio of two females to one male. However, during 1918 the male tanks would suddenly acquire a new importance when the Germans started to produce a tank of their own. After their initial shock in 1916, when the British first used their new weapon, the German high command declared that they were unimpressed with the allies unreliable, poorly armoured machines. If this was the sort of diabolical, underhanded weaponry the British were having to resort to they must be desperate and victory must soon come to the Imperial German forces. A German newspaper article from early 1918 said that *'when the first tanks appeared in the picture they were not to be underrated, but the German artillery soon dashed them to pieces. The inventor of the tank is Sir Tritton. He has probably through his invention, become a great help to England, but he will not bring victory to his war-lustful country or his invention'*. The Germans have never been a people known for alarmist, knee-jerk reactions and instead of panicking and rushing to create their own tanks they instead tried to find another answer to the new British weapon and put their faith and resources into new tactics, artillery and anti-tank technology. As Kitchener had prophesised when he first saw Mother at Hatfield, a well placed 77mm field gun was indeed the natural enemy of the tank, but these were not always readily available. Huge pits were dug as improvised tank traps and steel cables were strung along in front of the German trenches to deter tanks.

These methods took a great deal of time to prepare when what was really needed was a cheap, effective and portable anti-tank weapon that could be readily used by every German soldier at a moments notice. The Germans quickly discovered that standard rifle ammunition with the projectile head pulled off and re-inserted backwards would often punch through a tanks inadequate armour plate, whereas a standard pointed tip would usually deform and bounce off. This was actually an old trick used against enemy loop-holes and it worked reasonably well for tanks too until it was replaced by the 7.62mm 'K' round. The K round was used in the standard bolt action Mauser infantry rifle and each man at the front was issued with just ten of these specially developed armour-piercing bullets to be used only in the event of a tank attack. The new bullets proved to be very successful against the British tanks, but they were of a small calibre and in order to be affective, needed to be aimed very precisely at certain weak points on the advancing machine. Several other in the field methods were devised for despatching tanks including strapping a stick grenade to the side of a one gallon petrol tin, which was then thrown onto the roof of the passing tank. This was little more than a variation on the age old petrol bomb or *'Molotov cocktail'*. The grenade would explode, doing no particular harm to the tank, but the petrol tin would disintegrate leaving nothing but a cloud of burning petrol, which would rain down inside the tank through tiny gaps between the armour plate, the observation hatches and the exhaust ports making life very unpleasant for the crew inside. Another early anti-tank weapon was the multiple headed grenade, later known as the '*Geballte Ladung*'. Six stick grenade heads were removed from their handles and strapped around a single complete grenade. This would then be thrown into the path of an advancing tank and the combined explosive power of the seven grenades would hopefully be enough to blow one of its tracks off. A similar, but even more unsophisticated way of removing a tank track entailed simply stuffing as many grenades into a canvas bag as you could, then you would pull the pin on one of them and thrown the whole lot at the tank. None of these methods assured success, but perhaps the most obvious drawback with them is that the German with the petrol tin or bag of grenades had to be very close to his prey. Despite this major shortcoming all three of these improvised weapons were used again during the Second World War, not only by the German Army, but also by the British and by the Home Guard. As armour technology improved and plate thickness increased the reverse bullet method and the K round became less of a threat to tank crews, but in February 1918 Waffenfabrik Mauser at Oberdorf in Germany unveiled the 13.35mm (.50 Calibre) Mauser Tank Abwehr Gewehr. The new Mauser anti-tank rifle, also known as the M1918, was basically just a scaled up version of the standard 7.62mm M1898 infantry rifle. The M1918 fired a bullet four times heavier than the standard 7.62mm

Mid 1918 and the crew of a male Mk V take a break in the fighting to examine a lethal looking captured M1918 Mauser anti-tank rifle (The Tank Museum, Bovington)

ammunition, whilst still retaining a similar muzzle velocity. The inch long, half inch round, steel cored bullet would give a frightening 1inch (25mm) of armour penetration from almost 350ft (100m). It was the worlds very first, purpose built anti-tank rifle. It could be moved and operated by one man and before the armistice Mauser had produced over 15,000 examples. Although it was an excellent gun, in practice this huge and heavy piece of ordnance was understandably unpopular with the men who had to carry and fire it. It was over 5ft (1.5m) in length and including the bi-pod weighed almost 40 pounds (19Kg). The men tasked with firing the Mauser must have been incredibly strong as they would not only have to carry the huge anti-tank rifle around the battlefield, but they would very often also have to wear an uncomfortable steel breast-plate and the ridiculously heavy 'Stirnpanzer' which was a 6mm thick steel plate hooked onto the front of the standard German helmet. This extra armour was usually reserved for machine-gunners, sentries and snipers, but when the Mauser anti-tank rifle was added, these men became the elite '*Panzer Jager*' or Tank Hunters. The Mauser was fired in the standard way, from the shoulder whilst lying down in the prone position. It had no shock absorbers or dampening and the recoil as the massive gun was fired is unimaginable. In addition to its unwieldy size, weight and dangerous kick the Mauser M1918 was a bolt-action rifle with no magazine meaning that the breach would only hold one round at a time, consequently it was slow to operate and as soon as a shot had been fired the operators position was often betrayed and could soon be overrun by accompanying infantry or tanks. A few magazine equipped examples were produced, but it was too little and far too late to make any difference. The sheer weight of the Mauser meant that a retreating soldier was unlikely to want to carry the 40lb gun very far and would usually just remove the breach bolt and leave the gun itself behind. As a consequence many of these impressive rifles were captured almost intact and displayed to the public during fund raising events such as War Savings Certificate days.

As British and French tanks became more sophisticated and tank tactics improved, the Germans became more concerned and eventually they admitted that a final antidote to the tank menace would have to be found. To be exact if they wanted to defeat the Allied tanks and break their trenches they would need to think beyond anti-tank weapons and start producing tanks of their own. The Germans had actually been working on ideas for tracked fighting vehicles for almost as long as the British and the French. The German high command had pinned its hopes on one design in particular, a design that had been on the drawing board since 1916. Early in 1917, after much development work the Germans finally unveiled the A7V Sturmpanzerkraftwagen, and a truly awful creation it was too. The A7V was basically little more than a riveted steel box based on track set derived from the American Holt system and powered by a pair of 100 hp Daimler engines, very similar to the those used in the earlier marks of British tank. The finished A7V weighed over 30 tons and although it was of similar physical dimensions to the standard British Mk IV, it had a huge crew of eighteen men. The Sturmpanzerkraftwagen was fielded as both male and female variants, formidably armed either with a Belgian 57mm casement cannon and six MG 08/15 machine guns as a male or eight machine guns for the female variant. Unfortunately for the Germans, its considerable height, narrow track and poor weight distribution meant that it had an annoying tendency to fall over when crossing rough ground. The Germans only ever built 20 A7Vs, realising that the design was less than perfect. Instead they preferred to use captured British tanks and even went as far as imitating Walter Wilsons rhomboid design to create the A7V-U. This tank was tested in the summer of 1918, but once again, the German design had fundamental flaws, such as uneven weight distribution and poor visibility. Only one proto-type was completed and the A7V-U never went into mass production and never saw action during the war. This second unfortunate design made the Germans even more enthusiastic about using tried and tested

The German answer to the British tanks, the A7V Sturmpanzerkampfwagon. This example has fallen into French hands in late 1918

Its ability to deal with rough ground was just one of the A7Vs many design problems

Female Mk IV B30 'Bally Hackle II' was one of seven tanks put out of action by an anti-aircraft lorry at Fontaine Notre Dame during the Battle of Cambrai. The crew were lucky with only one of their number, 2nd Lieutenant Mercer being killed. The tank was subsequently retrieved and repaired by the Germans, note the large screw jacks and purpose built skis used during tank recovery operations

A ditched female Mk IV. Made by the British, captured and repainted by the Germans and now in French hands after the Battle of Pompelle

A medium Mk A Whippet goes through its paces for its new German owners after being captured intact

The A7VU, a poor imitation of the British rhomboid design (The Tank Museum, Bovington)

captured British tanks and they became very efficient in the retrieval and refurbishment of captured medium and heavy machines. Repainted, renamed and now known as *'Beutepanzers'* the tanks were soon put back into service against their previous owners. It is ironic that the country that appeared the least impressed and did the least to develop tanks during the First World War should lead the world in armoured vehicle development and tactics a mere 21 years later.

While Germany and England were concentrating on heavy tank design, in France they were working in the opposite direction by creating a small fast chaser type tank called the Renault FT-17. The little Renault was operated by just two men and although it was relatively thin skinned and poorly armed it become very popular with its crews, even becoming the favoured weapon of the newly formed American Tank Corps. The French did produce heavy tanks in the form of the St. Chamond and the Schneider CA1, both of which used the Holt track system as their basis and were consequently not very good on rough broken ground. It is the little FT-17 that will forever be seen as one of the best tanks produced by either side during the war and they proved their worth during Marshall Fochs great counter attack of July 1918. Like the British tanks and the German A7V, the little French Renault FT-17

The lightly armoured, but very capable female Renault FT17

The 33rd Division, Tank Corps of the Illinois National Guard take receipt of a brand new Ford made FT-17 and a specially adapted six wheeled lorry to carry it

also appeared as Madams and Monsieurs using the small 37mm cannon or a single machine-gun respectively. Although the Renault and even the ill-fated A7V both had their good points, in 1917 the British Mk IV had the battlefield to itself. The new mark was the finest tank in any army and in the hands of an experienced crew it was a revelation. The new tank was so good that the Tank Supply Committee in London received requests for complete Mark IV tanks from the French, Belgian and Russian governments. If these requests were to be answered tank production would have to go up to at least 150 new machines per week, a quite impossible number and so the tanks were never sent. Albert Stern thought that there was more to the Russian request than first met the eye as in his opinion the Russians had no need for tanks. In fact Stern was so suspicious of the Russian appeal that he concluded that it had actually come from the German General Staff pretending to be the Russians. The request was refused, but not long after a second request came though, supposedly from the Russians again, this time asking for the plans for the Mk IV. Stern thought this even more suspect as in his opinion the Russians had neither the resources nor the infrastructure to be able to build tanks. The requests kept coming, but in the end Stern says that no more was heard after they sent '*a childs drawing of a tank and incorrect details*'. It was no surprise that the Germans wanted to get their hands on the production details of the new Mk IV by any means possible, it was a vast improvement on previous designs and given the right ground and conditions it performed well in many of the wars most infamous battles. As the face of the war slowly began to change, so did the requirements that the tanks would need to fulfil and back in Lincoln, William Tritton began to think of the future.

William Fosters technical and commercial staff and three brand new Mk IV males taken during 'factory portrait day' in late 1917

CHAPTER III
Medium Tanks

The Medium A

Only one month after the first tank battle in September 1916 William Tritton and Albert Stern of the Landships Committee had met in London to discuss what would happen when the stalemate was broken and the war became a series of mobile cross-country battles. The general agreement was that the need then would be for small, fast, manoeuvrable tanks that could be driven by one man, negating the necessity for the gearsman. Immediately on his arrival back in Lincoln, Tritton and Rigby started work on the new design and by late February 1917 the new tank was finished. It is at this stage that Tritton should start to be referred to as Sir William Tritton as on February 13th 1917 he was granted a Knighthood by King George V. In order to mark the occasion of Trittons Knighthood the board and employees of Fosters

Now thrive the armourers as Whippets are hurriedly prepared in Lincoln for immediate despatch to the front

A tank in Action by John Hassall

commissioned an engraved silver rose bowl for the newly appointed Lady Tritton and an illuminated address and a painting to hang on the office wall of Sir William. The painting was by the renowned Royal Academy artist John Hassall, was titled '*A Tank in Action'* and according to the Louth Leader Newspaper depicted *'a Foster made machine riding triumphant across the battlefield with dawns early light illuminating its good works in vanquishing the Hun'*. Framed prints were created on stiff card and sold for £1 each in aid of the William Foster and Co Ltd Soldiers, Sailors and Airman's Comforts Fund. The tank shown in the painting is supposedly a Mk I tank in battle, but it has no steering gear at the back and actually has several features only even seen on the prototype tank, Mother. It is probable that Hassall used Mother as his model and that by the time of the sitting she had already been robbed of many parts for spares, probably including her tail. The current whereabouts of Lady Trittons silver rose bowl is unknown, but the original John Hassall painting is being held in storage at the Usher Art Gallery in Lincoln who have no plans to display it.

In March 1917 the new Sir William Tritton took his new tank to the Oldbury Trials. Perhaps the strangest looking machine amongst the weird and wonderful tanks on test at Oldbury was Fosters latest creation, known as the Tritton Chaser. The original Foster prototype had been stranger still with a cylindrical turret, just one machine gun mounting and the petrol tank hidden

A rare side view of the William Foster and Co Ltd prototype Whippet

Fosters new prototype Whippet taken from the front, note the single gun mounting

away at the rear of the machine. Trittons new tank was an amalgamation of ideas and features taken from Little Willie, the still born Flying Elephant and the French Renault FT-17. Trittons new tank would be a radical new machine in which he and Rigby would finally get the chance to try out several new theories they had been working on. The main problem they wanted to over come was the fact that tracked vehicles obviously lost a great deal of forward momentum when the vehicles brakes were used for steering. After several failed and very complicated designs Tritton decided to keep it simple and overcame the problem by using a system similar to the engine and gearbox design that he had created for the ill-fated Flying Elephant. The Tritton Chaser was driven by two separate engines, each one driving a single track. The new tank was steered using a steering wheel such as might be found in a standard road-going wheeled vehicle. However, instead of turning road wheels the Whippets steering wheel was used to increase and decrease the two throttles linked to it, making turning simply a matter of decreasing the throttle on one engine whilst increasing the revs on the other in one smooth rotary movement with no need to apply the brakes. It may be assumed that if one engine was drastically out of tune or a tank lost an engine completely then it would simply go around in circles, but Tritton and Rigby had considered this. The Whippet had a clever system including a slip clutch that sat between the two final output drives. This clutch would slip if the power output differed between the two sides by more than 12hp, so if one engine failed you could still carry on moving by using very low revs on the remaining engine which would run both tracks at a very slow limp via the slip clutch. Turning was another matter, if only the left hand engine was still running, then turning right would be no problem but to turn left you would have to turn right through 270 degrees. Little Willie, Mother and the Mark I machines were all equipped with steering wheels, but unlike the Whippet the steering wheel in these machines was there just to operate the cables connected to the tail gear, which in reality had little affect on the machines direction of travel. Despite the ingenious slip clutch, from an engineering point of view the Tritton Chasers twin-engine layout and steering seems quite archaic by modern standards, but the system solved the problem of being able to steer a tank whilst still travelling forwards without experiencing any loss in speed. The engines chosen for the new machine were a pair of 45hp four-cylinder Tylors, being a tried and tested unit used in lorries and buses of the time. Although steering the Tritton Chaser was simple enough in theory, the twin-engined layout presented the driver with other challenges in that if you have two engines, you have two sets of controls. You would be driving whist trying to operate two gear selectors and two clutches, all whist steering and looking out for shell-holes, no wonder that it was often said that the Tank Corps were recruiting circus jugglers to drive the new tank.

In skilled hands, a well-tuned Tritton Chaser was all that had been hoped for. It was a fraction of the weight of a Mk IV at around 12 tons (12,200kg) and could do almost twice the speed at a staggering 7.5 mph (12kph) on level

A brand new Whippet gets the star treatment in Fosters Yard

From Lincoln to Russia, a Whippet and its new owners in the mid 1920s

ground. It carried four .303 calibre Hotchkiss machine guns, one on each side of its static turret. Originally it was planned to carry a crew of four, a driver, commander and two machine gunners. However, in practice, the cramped conditions inside the cab dictated that the crew was usually cut to three with the commander often doubling as a gunner.

In spite of its noteworthy appearance at Oldbury the Tritton Chaser was not scheduled for production as the trenches were yet to be broken there was no need for a chaser tank when there was nothing to chase. The War Office was committed to heavy tanks for the time being, but just two months later in June 1917 William Foster and Co LTD received an order to build 200 of the new design, which had now been given the official title of the Medium Mark A. Fosters began work on the order immediately and in September 1917 Major Phillip Johnson, the Tank Corps workshops officer and technical advisor, visited the factory in Lincoln in order to inspect the new tank. Johnson had started at the bottom and worked his way up, from driver, he had quickly risen through the ranks to become workshops officer and as such knew tanks, their tactics and driving inside out. It was with his assistance that the Central Workshops in France had designed many 'in the field' improvements to the tank, such as Spuds and Unditching Gear. Spuds were simply individual bolt on track extensions that acted like snowshoes for the rhomboid tanks, spreading the machines weight and giving the tracks extra grip in muddy conditions. The spuds looked like pressed steel anchors, attached at intervals around the track length and were based on an old idea used by traction engine manufacturers and had perhaps even been suggested by William Tritton. Spuds were also specially designed for the Whippet, but instead of spreading the machines weight the wood and steel shoes were mainly intended to give extra grip and can often be seen hanging from the back of Whippets when not in use. The Unditching Gear was another splendid idea and consisted basically of a huge, iron bound oak beam carried across rails running along the top of the tanks hull. If the tank became stuck the beam could be attached to the tracks where it would travel along the rails and around the entire hull. When the beam came into contact with the ground it would, hopefully, give enough grip to get the machine moving again. As the beam and the rails on which it travelled were only ever fitted by Central Workshops in France, their presence on the back of a tank is always a good indication of whether it actually saw action in France or spent its time as a training machine in England. The unditching beam and gear sounds like an obvious solution to a difficult problem, but shows an inspirational grasp of simple, yet effective engineering that would become the trademark of Central Workshops and Phillip Johnson. When Johnson visited Lincoln in 1917 and saw Trittons prototype medium tank, which had by now become known as the Whippet due to its comparatively high speed, he announced that he was impressed. The prototype Whippet had done an unbelievable 870 miles (1400km) during trials and testing without a single mechanical breakdown. The only problem with the new machine was the 45hp engines. Tritton had hoped to produce a tank that could do in excess of 10mph (14kph) and would have preferred to

A Whippet takes a break to help launch a French fishing boat

have used more powerful engines for the Medium A, but his hands were tied by War Office specifications and material shortages. After the war he rather unkindly said of the tank that '*the choice of the 45 hp engines had spoiled an otherwise useful machine'*. Johnson must have been thinking along similar lines, but he was looking for a bit more than a stately 10mph. In late 1918 he adapted a standard production Whippet by giving it sprung tracks and a single 360hp V12 Rolls-Royce Eagle aero engine instead of the twin Tylors. Johnson had been working on the idea that effective suspension and more horse power were the two main factors in gaining increased cross-country speed and with his modified Whippet he had proved it. The sprung tracks and an extra 270hp made a machine that was easily capable of 30mph (48kph). Unfortunately, Johnsons system was expensive to produce and with the war drawing to a close the prototype never went into production.

More speed would have been a great advantage, but even so the standard Medium A Whippet had an impressive war record. It's speed and agility gained it the title of 'the new Cavalry'. This can hardly have won the tanks or their crews many friends in the 'old school' as right from the beginning of mechanisation the Cavalry and many in the military top brass had opposed any step forward in mechanisation. Much of the problem they had with mechanised weapons such as tanks stemmed purely from the fact that things

Phillips Johnsons magnificent Rolls-Royce engined 30mph Whippet (The Tank Museum, Bovington)

just were not done that way and that the Cavalry was symbolic of the way that Gentlemen should conduct themselves during a war. The way many people thought is well summed up in a Punch cartoon of the time in which the Cavalry is described as being there to *'give tone to what would otherwise be a mere vulgar brawl'*. Despite its lowly image the new Cavalry of Whippets saw service for the first time at the Mailly-Maillet on 26th March 1918 and the new tanks contribution to the war soon began to speak for itself. The Whippet was such a good machine that they were specifically requested for use in the renewed desert fighting in Palestine, although no machines were actually delivered in time. The role of the chaser tank could no longer be ignored and the new cavalry was here to stay.

The Medium B

In a surprisingly logical step the military decided that the next tank after the Medium A would be the Medium B. However, in a total break from tradition the new tank would be designed entirely by Walter Wilson. In September 1918 the Metropolitan Company in Birmingham unveiled the prototype and it immediately became apparent that the new tank had several teething problems and one serious flaw. The transportation of tanks by rail had always been a major headache and with the new design Wilson meant to address this. He made the Medium B as short as possible with the aim that the new machine would be able to fit easily onto any standard flat railway truck. Unfortunately, making the tank so very short meant that interior space was in short supply and the engine compartment in particular was incredibly cramped. To make matters worse the engine bay could not be accessed

Walter Wilson exits his ill-fated Medium B (The Tank Museum, Bovington)

inside the tank, not a problem when doing maintenance at the depot, but even a simple mechanical fault in battle would have entailed one of the crew getting out under fire to attend to the problem. In many other aspects Wilsons machine was excellent, but only 80 of the proposed order for 700 Medium B tanks were produced before the order was cancelled. Several were apparently used by the 17th Armoured Car Division in 1918, during the troubles in Ireland, but that was as near as the Medium B ever got to combat. None of Wilsons Medium Bs were built by William Foster and Co Ltd at Lincoln and unfortunately, none of the ill-fated tanks are known to have survived anywhere in the world.

The Medium C

In the William Foster and Co engine records for late 1917, early 1918, there are several entries for engines to be used in a new tank, described as the Mk VI. At first glance this appears odd as the MK VI was a heavy tank that had already been proposed in June 1917. After the wooden mock-up had been built by the Metropolitan Company the production of the tank was not approved due to its radical departure from the existing design and the Tank Supply Committees production commitments to the new heavy MK V. It is not until one sees the word 'Hornet' on the next page that all becomes clear. The Hornet was to be a replacement for Wilsons ill-fated Medium B and was officially designated the Medium C. The new tank was another design from the drawing board of Tritton and Rigby. The Medium B and C were the first heavy tanks to be designed separately by Wilson and Tritton and

The future of tank design, the prototype Hornet stands proudly on William Fosters testing ground

The old gives way to the new as a fleet of Medium C Tanks are paraded at Aldershot in 1923

apparently there was a great deal of good-humoured rivalry between the two men. Although the Hornet entered service just too late to see action during the Great War, it is generally regarded as the best tank of the era. It combined the experience and technological advances gained during production of all previous marks of tanks in one package. For the first time the conditions under which the crew had to work were taken into consideration. Unlike Wilsons design the fighting compartment of the Medium C was completely isolated, but still accessible from the well ventilated engine compartment which housed the same six cylinder 150hp Ricardo petrol engine used in the Mk V machine. This partition lowered the noise and pollution levels for the crew, but this design consideration was not just for the crews personal comfort, it was a serious safety measure. Tank Corps losses were running at an incredible 40% and it is an often quoted fact that more men were killed

The finest tank of the period, a Lincoln made Medium C Hornet performs for the crowds at official testing in 1918 (The Tank Museum, Bovington)

by carbon monoxide poisoning in the early tanks than were actually killed in battle. There were also voice tubes connecting each of the positions, negating the time honoured necessity to bang out coded instructions on the inside of the hull with a large spanner. The Hornet could be driven by just one man and the commander had a small rotating cupola at the rear of the turret from where he had a good all round view of the battlefield. As with the first rhomboid tanks the Hornet was unsprung and the tracks ran all around the hull, and on good ground the improved power plant meant that it could even outrun a Whippet, but only just, the Hornet had a top speed of 8mph (12.8kph). The turret had provision for five machine guns, though only four were fitted. The spare mounting took the form of a hatch in the roof, one of the Hotchkiss machine guns could be moved onto the mounting beneath the hatch for use as an anti-aircraft gun, this was very useful piece of design in theory, but in practice it left the gunner completely exposed to enemy fire. No male version of the Hornet was ever put into production, although it had originally been intended to make both variants. Following the Hornets successful trials an order was placed for 200. In October 1918 a further order for 4,000 female and 2,000 male machines was received. This was immediately cancelled due to the old problem of production shortages. At a Royal Commission hearing after the war Tritton was asked why the Medium C had taken so long to come into service, he answered that '*many months had been lost on tank construction because of the failure of the Mechanical Warfare Supply Department to deliver parts. The ministry had promised gearboxes and engines, but the delivery fell further and further behind until the delay was well over a*

Medium C 'Princess Mary' earns her keep during a military show in the early 1920s (The Tank Museum, Bovington)

A brace of Hornets race for the assembled masses at the 1923 Royal Tournament

year'. Perhaps just as pertinent as the supply shortages was the fact that by now everyone could see that the end of the war was in sight. Several other engineering manufacturers were given orders to begin production of Medium C Hornet machines, but shortly after the end of hostilities all contracts were cancelled leaving William Fosters as the sole contractor working on the new tank. By February 1919 the company had delivered 36 of the original 200 tanks, but soon afterwards the order for the remaining 164 units was cancelled. The Medium C was a significant milestone in tank design and development, it can rightfully be seen as the direct link between the wartime rhomboids and the more conventional designs that followed it. The last of the 36 Foster built Hornets built went out of service in late 1923. No Medium C Hornets are known to survive anywhere in the world today.

CHAPTER IV

The Heavy Tanks

With the stalemate of the trenches now broken medium tanks like the Whippet and the Renault FT-17 were seen as the future of armoured warfare. Other marks of heavy tank followed the Mk IV and although many radical features were tested on the following marks none of the new machines differed greatly from earlier tanks and none of them were produced by Lincoln firms. An interesting variant of the Mk IV was produced in Lincoln in 1917, it was created as an answer to the wider trenches that the Germans had begun to dig and it was curiously called the Tadpole. At the Battle of Cambrai in November 1917 the supposedly uncrossable trenches

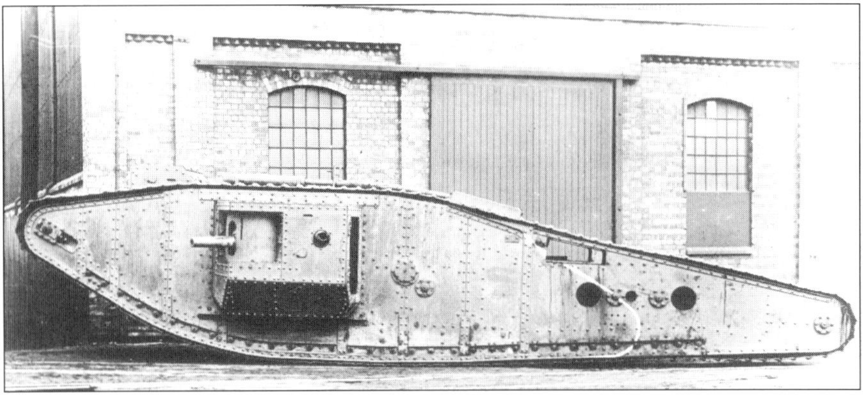

The prototype Mk IV male Tadpole in Fosters works yard. The white line shows the profile of a standard Mk IV machine (The Tank Museum, Bovington)

of the Hindenburg line had been breached by using fasines. The fasine was an eight-foot tall (2.5m) bale of brushwood bound together with steel cables and dropped into the trench by the first tank. There were several fundamental problems with the design of the fascine; it was difficult and dangerous to produce, if either of the cables snapped anyone standing nearby would surely be cut in two. The fascines were made by men of the Chinese Labour Corps and finding enough brushwood to produce fascines for the entire Tank Corps led to the deforestation of a large part of Northern France. In battle, the fascines were often shot away or more often than not, set on fire. Added to these drawbacks was the fact that the mechanism used to deploy them often jammed leaving no alternative than to get out of the tank under enemy fire and drop the fascine by hand. It was decided that the answer to the broader German trenches would be a longer British tank. Fosters created a longer tank by attaching extended rear horns onto a standard male Mk IV increasing the machines overall length by over 13ft (4m), thus making it capable of crossing a much wider trench. This adaptation of the standard machine became known as the Tadpole, due to it's long tail, and would appear to be an obvious and simply produced solution to the problem of wide trenches. Indeed it would have been if had not been for the L over C ratio. When Tritton and Wilson designed Little Willie, Mother and eventually the Mk I they had no idea that the length of the track in contact with the ground, L, is directly linked to the distance between the track centres or C. The longer the length of the tracks footprint on the ground, the wider the distance between the tracks has to be or it will become increasingly more difficult to steer the finished machine. Today this equation is a fundamental of tracked vehicle design, but in 1916 nobody had even considered it. The L over C ratio for Mother and the Mk I was dangerously close to the limit, but with the creation of the Tadpole, Rigby and Tritton had gone too far. The machine was very slow to steer and the length of the rear horns made them structurally weak with a tendency to twist. The Tadpole was tested at Burton Park and on Fosters own testing ground at Lincoln pulling fully loaded supply sledges. The finished machine even had a mortar fitted between the rear horns in order that a whole squadron

The yard at Tank Corps Central Workshops in France. In the foreground is a huge store of Tadpole Tail units, made in Lincoln, shipped to France and left to rust (Bill Baker)

A female Tadpole shows off during trials in 1917 (The Tank Museum, Bovington)

of advancing Tadpoles could lay down their own creeping barrage. Although the design was approved and batches of Tadpole conversion kits were sent out to the Tank Corps Central Workshops in France there is little evidence to suggest they were ever even fitted and no evidence at all that they were used in combat. Central Workshops thought they could do better and attempted to solve the problem themselves by taking the more obvious route of simply slicing a standard Mk IV in half behind the sponsons and adding another 6ft (1.8m) to its length. Although more structurally sound than the Tadpole their creation still suffered from the same steering and control problems.

An ex British Mk V in service with the Russian Tank Corps in 1921

Eventually the L over C ratio was solved and the longer and wider Mk V was introduced. Externally the Mk V looked like an extended version of the Mk IV, but internally it had two very important differences. The front arrangement of vision flaps and Lewis gun port was revised in order to partly address the old problem raised by the Tank Corps in 1917. With the introduction of the Mk V for the first time just one man could drive the tank. This was due to the introduction of Wilsons epicyclic transmission that had first been tested at the Oldbury trials in 1917. Although no Mk V tanks were produced in Lincoln, many of the design drawings came from William Rigbys drawing board and William Tritton can even claim some small credit for its initial production and success.

Motoring mastermind, Sir Harry Ricardo (The Tank Museum, Bovington)

As soon as the tank had emerged in 1916 it had been patently obvious that no matter how good the Daimler 105hp engine was for it's intended role, it was no way near powerful enough to propel tanks. The engine had been pushed to its very limits trying to move almost 30 tons (30,490kg) of armoured vehicle through muddy ground, but even so the Daimler unit was the only engine even close to being suitable for the job. Everyone was in agreement that a completely new design of more powerful engine was urgently needed, but every manufacturer capable of taking on such a task was already at full stretch making aeroplane, lorry and marine engines. The answer came quite by accident when William Tritton remembered his pre-war dealings with a young engineer by the name of Harry Ricardo. Tritton went to Ricardo in spring 1916 to advise him on the feasibility of producing a larger purpose built power unit. Tritton had chosen his man well for as Ricardo put it himself *'It had been my good fortune to have been on intimate terms with the petrol engine since its first appearance in England'*. Ricardo had almost become involved with tanks right at the start in late 1915 when he had been approached to create a system of hydraulic jacks that could be used to inch a tank into position once it had been driven onto a railway truck. However, as tank crews became more adept at moving their machines such a device was no longer needed and the design was dropped.

In mid 1916 Ricardo was made an honorary member of 20 Squadron RNAS and was soon introduced to the tank and its Daimler power plant. He was greatly impressed by the tanks, but utterly horrified by the unsuitable engine and resolved to do something about it. Even though it was the most powerful automobile petrol engine in production the Daimler unit was desperately under-powered for use in tanks. Perhaps even worse was the fact that the engines entire internal lubrication system was totally wrong for the job it was being made to do. The engine oil settled in the sump pan and the big ends and crankshaft were bathed in it. This was fine on flat roads, but when the tank was at an angle all of the engine oil would slosh to one end of the sump and the big ends at the other end would be operating dry. It would not take long before the engine of a tank trying to extricate itself from a shell-hole at high revs would become completely seized, and they often did. The sleeve-valve system gave no means by which the oil could be spread circumferentially and only by pouring oil over the outer sleeve could the engine be prevented from seizing. This was not only inconvenient, it was downright dangerous as the oil would inevitably find its way into the red hot exhaust ports and create great clouds of blue smoke, meaning that very often the crew could not see the enemy, but the enemy could not miss the approaching tank. Ricardo was asked to find a suitable manufacturer to create an engine that would overcome these problems. He toured all of the main engine makers such as Rolls-Royce, Sunbeam and Daimler and got the same answer each time, they could easily make the new engine, but could not spare the time, men or resources to design and test it. Ricardo reported back to Albert Stern and Tennyson d'Eyncourt. Things looked bleak until d'Eyncourt asked Ricardo if he thought he could design the engine himself. This was exactly what Ricardo had been hoping for and he agreed to try his hand immediately. Ricardo returned to his Grandfathers engineering firm and got to work. The new engine was originally intended to be 200hp, but later tests showed that this would be too much power and 150hp would be more than sufficient. The 150hp engine would be coupled to the revolutionary one man operated epicyclic transmission designed by Walter Wilson. The internal lubrication of the new engine was excellent at any angle and Ricardo had even taken into account the poor quality of petrol being supplied to the Army in France and made provisions for his engine to run perfectly well on the substandard low octane fuel. When the design was finished seven engineering companies were approached to produce it. It was proposed that two or three hundred units would be a realistic number for the firms to produce in total, but by the end of the war between them they had built almost 9000. Many of the new units were produced by Richard Hornsby and Sons of Grantham who had been the first European manufacturer to produce a viable chain track machine powered by an internal combustion engine in 1906. The Ricardo designed engine was originally intended for the Mk IV, but the cost, both financial and in time, became increasingly prohibitive and the new engine and transmission were not ready until the introduction of the Mk V. Despite a few teething problems and higher interior working temperatures from the larger engine, the tank became an infinitely superior machine overnight.

In addition to the new power plant the Mk V had wider tracks, an improved hull layout, better ventilation and rear facing machine-guns. Over the coming months the Mk V was gradually lengthened until it eventually became the huge Mk V** that was capable of both fighting and carrying troops. The main problem with widening tanks was that the railway still remained at a standard gauge, so instead of expanding outwards the tracks had to be widened inwards, making the hull narrower. However, due to one man operation, purpose built engines and more room lengthways the evolving new tanks did not become noticeably more cramped inside the fighting compartment than the old Mk I had been. As previously mentioned there was no Mk VI tank, but there was a Mk VII. Once again the Mk VII was an old fashioned rhomboid design similar to the Mk I, IV and V, but the interior of the new machine was full of state of the art technology. The tank was equipped with a complicated hydraulic drive system that drove the tracks through a pair of variable speed gear pumps directly coupled to the engine. Another innovation seen on the Mk VII was that it was the first tank to be fitted with an electric starter. All the previous marks of tank needed at least four men to crank the engine over by hand and there is one recorded incidence of the engine kicking back and killing one of the crew. Despite the tanks innovative features, only three Mk VIIs were ever produced, all by Brown Brothers of Edinburgh. The next machine after the Mk VII was conveniently called the Mk VIII. Although along similar lines the tank was quite a digression from the previous tank designs. It had a huge ovoid shaped hull and was powered by a massive 300hp V12 aeroplane derived engine. The Mk VIII came as a result of the Anglo-American Treaty signed in November 1917 and was a joint design created between the two countries. The finished machine, although designated the Mk VIII, became known as the Liberty Tank or the International. The Mk VIII was an excellent tank and in tried and tested American fashion it bristled with armourment. It was equipped with a pair of six-pound guns and so many machine guns, seven in fact, that it could not be decided weather the new machine should be designated male or female. The

The imposing form of the American Mk VIII Liberty tank in 1918

armour plating of the hull was uprated to a maximum of over half an inch (16mm) thick and there was so much space inside the monster tank that it could carry its standard crew of 12 men plus another 20 passengers. The Mk VIII was a splendid design and a potent demonstration of engineering skill and trust between America and Great Britain.

Even though there had been plans for a Mk X, which would have basically been an improved version of the Mk V, Britain's last heavy tank to be produced before the armistice was the Mk IX. Appropriately enough the story of the heavy tanks of the First World War finishes where it began, in Lincolnshire. The MK IX was to be a specially created machine, not just a modified version of the MK I. The new tank was originally designed as a multi-role vehicle, capable of fighting, carrying supplies or transporting troops. Two prototypes were built by Armstrong and Whitworths in Newcastle-upon-Tyne. In trials the new machine impressed everyone and an order for 200 was placed with the Gainsborough based steam traction engine manufacturers, Marshall Sons and Company. Marshalls foundry at the Britannia Ironworks was very much like William Fosters, it was a relatively small Lincolnshire based concern producing high quality steam engines for the home market and the empire. In order to maximise the internal dimensions of the Mk IX the outer track frames were discarded, this would have left the tank in danger of collapsing under its own weight, so steel girders were added transversely across the floor. The 150hp Ricardo engine was brought forward as far as possible and

The answer to the problem of supply, Gainsborough's Mk IX combined mobile stores carrier and armoured fighting vehicle (The Tank Museum, Bovington)

the gearbox was placed as far back as it could be, leaving the entire centre section empty apart from a propeller shaft running along the floor. The Mk IX was a completely redesigned shape with a steeply sloping square front end, leading to its nickname of The Pig. The tank had no sponsons and only two machine guns for armourment, one at the front and one at the rear. Instead of sponsons the Mk IX had two large oval doors on either side and could hold thirty fully kitted soldiers or 10 tons (1020kg) of supplies. Even though the machine did not have any side facing armourment it did have eight loopholes down each side, so if need be, its passengers could defend it from within. At least one Mk IX was converted into a fully equipped armoured ambulance and another became the worlds first amphibious tank. By strapping oil drums along both sides and sealing up all the doors, loopholes and observation ports the tank became almost watertight. A conning tower was added on the roof from which the machine could be operated by means of extended rods linked to the internal controls. Metal flaps were bolted onto the tracks which would lay flat when the tank was being driven on dry land, but when the tank was in the water they would hang down and work against the water like the wheels of a paddle steamer. The amphibious Mk IX was tested in the Welsh Harp and worked quite satisfactorily although there are reports that it leaked so badly that it had to be abandoned, eventually sinking never to be seen again. Of the original order of 200, Marshalls delivered just 35 before the end of hostilities and another 11 before the end of production in March 1919. If you don't count the amphibious version supposedly at the bottom of the Welsh Harp, only one example of the 46 Gainsborough made MK IX supply tanks still survives, it can be seen at the Tank Museum at Bovington.

The worlds first amphibious tank, the swimming Mk IX takes to the water for the first time in The Welsh Harp (The Tank Museum, Bovington)

In addition to the principal marks of heavy and medium tanks numerous experimental machines and different variants were produced. Tanks were adapted for many differing roles including mine clearance, heavy lifting, bridge laying and supply. Getting supplies to the fighting man at the front without having them captured or blown up by the opposing side has been a major problem throughout the history of warfare. With the advent of the tank, supply suddenly became less of an obstacle. Originally, worn out or battle damaged tanks had their sponsons plated over and their hulls filled with stores, eventually the Mk IX purpose built supply tank evolved, but in the years between the two ideas there were numerous attempts to solve the supply problem. Massive sledges piled high with stores were towed by tanks, sometimes two or three being pulled by the same tank. However, the idea of sledges was soon dropped due to their vulnerability to enemy fire and their tendency to dig into rough ground, thus ending up carrying as much mud as they did supplies. Something better was clearly required and for this the War Office returned to Lincoln, not to Fosters this time, instead to Ruston and Proctor and Co LTD. The firm of Ruston and Proctors had been created when Joseph Ruston became a partner in Burton and Proctors Millwrights of Lincoln. The company became Ruston and Proctor and Co LTD and by 1914 was employing over 5000 people. The range of items made by Rustons during the war is truly mind boggling. They seem to have made everything from bombs, poison gas machines, flame throwers, lorries, submarine engines to a machine for the Ticklers Jam factory that generated gas for lighting from old plum stones. Rustons became famous during the First World War for building aeroplanes such as the Sopwith Strutter, the Bristol B.E.2c and the Sopwith

One of the many experimental machines developed during the war. This old Mk II has been fitted with a jib and an electro magnet apparently for bomb disposal work

Camel, but were even concerned with the tank story as they even made complete sponsons and cataract pumps for tanks to be delivered for fitment at Fosters. When they were approached regarding the problems of supply at the front, they came up with the Ruston Stores Roller. Their creation was a huge trailer that looked like a road roller that would be towed behind a tank.

The Huge and quite impractical Ruston Stores Roller

Brand new sponsons in Rustons workshops awaiting delivery across Lincoln to Fosters works

The outer casing rolled along the road whilst an enormous armoured box filled with stores and suspended from the main axle stayed upright inside due to a large concrete filled weight at the bottom. The stores were accessed via sliding doors in the side of the internal box. The Rustons Stores Roller was a practicable solution to the problem of supply, although as it would have been taller than the tank that was towing it, the roller would have presented an unmissable target to the German guns. The question of its massive profile on the battlefield became a purely academic one as by the time it was built specially created MK IX supply tanks were being designed and the roller never progressed beyond the prototype stage.

With the end of hostilities the problem of supply gave way to one of peacekeeping. Post-war Germany was in turmoil and a full peace treaty was still to be signed. Genuine fears that the war might resume before peace could officially be signed resulted in the creation of the first tank made specially for urban warfare. This hybrid was created by combining the destructive power of the 6Pdr equipped male with the anti-personnel capabilities of the female tank, resulting in the hermaphrodite tank. In reality the hermaphrodites were simply standard tanks, usually Mk Vs, with the sponsons swapped about so that they had the best of both worlds by having with a single 6Pdr in one side and machine guns on the opposite side. Three examples of the hermaphrodite tank still survive, all in Russia. One example can be seen at the Kubinka Tank Institute and two more are on permanent display in the City of Lugansk.

Mk V hermaphrodites on peacekeeping duties in post-war Germany 1919

With the introduction of each new variant and the tanks eventual successes, such as the battles of Cambrai, Hamel and Amiens, it was becoming clear that the new weapon was here to stay. Those who had fought against red tape and narrow minds, had triumphed in pioneering its development and fighting techniques. The workers at Fosters and the other tank factories had been vindicated, the gamble had paid off and the tank was shown to be the future face of warfare.

A splendid photograph taken from the Tank Corps Christmas card for 1918 that shows their entire arsenal including a Mark V, a Rolls-Royce armoured car and a Medium C

CHAPTER V
For Valour

On the 29th January 1856 a new gallantry medal was created by Royal Warrant and on 26th June 1857 Queen Victoria awarded the medal for the first time. It was retroactive to cover the Crimean War of 1854 and it was to be known as the Victoria Cross. The VC, as it was officially abbreviated, was, and still is, the highest decoration Britain could give. When it was first unveiled the unassuming new medal was immediately ridiculed by many of the press and the military alike. Unlike other prestigious decorations it had no gold, jewels or ostentatious enamelling. It was a simple, stylish bronze cross with a dark maroon ribbon, the front face bore a lion and crown and the words 'For Valour'. The reverse would be engraved with details of the recipient, his unit and the date of the action for which it was awarded. The bronze for the new medal is reputed to have been taken from two Chinese cannons that were taken from the Russians during the siege of Sevastopol which came to an end in 1855, although recently some historians have begun to shed doubts on this story.

Since its instigation in 1856 a total of 1,356 Victoria Crosses have been awarded. It is arguably the worlds most famous gallantry medal and as such has a fascinating history and mythology of it's own. Although the medal can only be awarded to British and Commonwealth nations, in 1919 the American unknown soldier was awarded the VC as a symbolic gesture to thank America and show how much their help and bravery was appreciated by the allies, in return Britain's Unknown Warrior was awarded America's highest honour.

Of all the regiments eligible for the Victoria Cross, perhaps rather ironically it is the Royal Army Medical Corps who have been awarded the most with a staggering total of 31. The Tank Corps did not come into being until 1916, but remarkably they succeeded in winning 4 VCs before the armistice in 1918. The four Great War Tank Corps VCs were all posthumously awarded and all four of them were won either directly by the crews of Lincoln made tanks or by small detachments that included Lincoln machines.

By the winter of 1917 the fighting around Zonnebeke in Belgium had reached it's pinnacle. It was part of the Third Battle of Ypres and the stalemate was almost total. Knee deep mud, rats as big as cats and piles of dead soldiers of both sides scattered in no-mans land. The fighting here had become the stereotypical image of trench warfare, mobility had been lost and the Tank was seen as the way forwards. On the 4th October 1917 they moved in and on this day Captain Clement Robertson would win the Tank Corps it's first Victoria Cross.

Captian Clement Robertson

Clement Robertson was born on the 15th December 1890 in Pietermaritzburg in South Africa. His father, an officer in the Royal Artillery, was serving there at the time. As soon as he was old enough to travel he and his mother returned to the family home in Delgany, Ireland. Young Clement went to school at Haileybury where he proved to be an intelligent boy. After school he went on to further education at Trinity College, Dublin. When not studying Clement enjoyed sports and was one of the founding members of the Delgany Golf Club in County Wicklow. When war was declared he soon joined up and by late 1916 he was serving in the Queens Royal West Surreys as an acting Captain, on special reserve to the newly formed Tank Corps.

On the 4th October 1917 he found himself in the thick of the fighting at Zonnebeke. The tanks would usually leave their starting points before first light and would be travelling in the dark over broken, shell torn ground. It was standard practice for recce parties to go into no-mans land prior to a tank battle to map the ground and lay white tape to mark the route that should be followed during the advance. From 30th September to the 4th October Captain Robertson and his batman had spent every night doing just this under heavy fire around Polygon Wood. At six o'clock in the morning Captain Robertson had led his four tanks to their starting points ready to go into action against the Germans entrenched at Reutel, but overnight shell fire had destroyed large sections of the white marker tape along a 500 yard stretch of the route. Robertson realised that without the tape the tank crews would become lost, so without any thought for himself he left his tank and continued on foot, guiding the tanks around obstacles and shell holes until they reached their objectives. During the entire time he was in the open he was under heavy shell, machine-gun and rifle fire. The official citation for Clement Robertsons VC states that '*although knowing that this action would inevitably cost him his life, Capt. Robertson deliberately continued to lead his tanks*'. Unfortunately the citation was all too correct as just as Captain Robertson and the tanks reached their objective he was shot through the head and killed. Captain Clement Robertson was 27 years old when he earned his postumous VC, he is buried at Oxford Road Cemetery in Belgium.

Captain Robertson had gained his VC in the winter fighting around the Ypres salient where wounded men could drown in shell holes and tanks disappeared up to their sponsons in the thick stinking black mud. The new machine had been fielded in too small numbers and was not suited to the conditions. Those who doubted the new weapon seized upon it as a farcical waste of manpower and resources, some even said that the Tank Corps had been sent to Ypres in a deliberate attempt to discredit them, but just one month later they would get the chance they had been waiting for, a chance to prove themselves at last.

On the 20th November 1917 the Tank Corps trundled into the history books as the most famous tank engagement of the Great War began. The offensive was officially known as Operation GY but the Battle of Cambrai as it became is now the stuff of legends. It is often wrongly quoted as the first time that tanks were used in battle and collectors will pay double or even triple the going rate for medals and memorabilia relating to this one engagement alone.

Cambrai was a quiet part of the front, soldiers from both sides would be posted there for a rest from the fighting with the area eventually becoming known to the Germans as the 'Flanders Sanatorium'. Due to this the ground was open, mostly unbroken by shell fire and perfect for tank warfare. On the day before the battle Colonel Hugh Elles issued his famous 'Special Order No. 6' to the men of the Tank Corps. The order was a hand written, personal message from Elles to the men of the Corps and said that '*Tomorrow the Tank Corps will have the chance for which it has been waiting for many months, to operate on good ground in the van of the battle in the light of past experience I leave the good name of the Tank Corps, with great confidence, in their hands*'. Order No. 6 was to be 'sexed-up' by the press when it was later shortened to '*England expects every tank to do it's damnedest*'. Elles was much more than a figurehead for the corps as when the battle opened at 6.20am on the 20th November 1917 he actually went into battle with his men in the curiously named male Mark IV 'Hilda'. With such leadership and so much to prove it is no wonder that the Battle of Cambrai was such a triumph for the Tank Corps. However, as Colonel Mark Dillon MC would later say of the battle '*a tank is very good for taking a position, but is not so good for holding it'*. The gains made by the tanks at Cambrai were not exploited by the cavalry and after a German counter-attack the battle ended in a near draw.

Despite the political infighting and the disappointing end to the promising start for Cambrai, the Tank Corps showed that given the right conditions the tank was a valuable asset to have on the battlefield. The fighting around Ypres had shown what a tank could not do, but the Battle of Cambrai had now shown exactly what tanks could do and it showed that the men who drove them were beyond reproach. It was perhaps inevitable that it was at Cambrai that the corps would gain their second Victoria Cross.

Richard William Leslie Wain first went to Llandaff Cathedral School and then on to the prestigious St. Bees in Cumbria. He had a brilliant mind and in early 1914 he gained a place at Oxford. Unfortunately, any scholarly aspirations he had would have to wait and he immediately volunteered in the Public Schools Battalion. From there he was given a commission in the Manchester

Captain Richard William Leslie Wain

Regiment and took part in the battle of the Somme, actually being one of the few survivors of the infamous Manchesters charge on the 1st July 1916. Although he survived 'going over the top' he was badly wounded and was sent back to England on sick leave. Whilst at home he visited his old school of St. Bees, where he presented a talk to the boys and gave trophies at the annual prize giving. The school had a remarkable record, with old boys winning no less than three Victoria Crosses during the First World War. By coincidence one of the other St. Bees VC recipients was William Leefe Robinson who received his medal in 1916 when he became the first pilot to shoot down a Zeppelin over British soil, the aircraft he was piloting was a B.E.2.c No. 2673 produced by Rustons in Lincoln.

After he had been declared fit for active service Captain Wain returned to the army, but by now he had requested and had been accepted for transfer to the Tank Corps. After his training in England and posting to France he was just in time to join in with the Tank Corps biggest and most famous exploit of the war, the Battle of Cambrai. Captain Wain was in command of the tank section of the 1st Company, 'A' battalion and went into battle on board tank A2 Abou-Ben-Adam II commanded by Lt. C.W. Duncan. The tank bore the production number 2399 and engine number 14396M showing that it was the second to last Mark IV produced by William Foster and Co Ltd in August-September 1917. When Wain set out aboard his Lincoln made machine he cannot have realised that it would be his last morning on earth and that by the end of the day he would have earned a Victoria Cross for one of the most remarkable feats of heroism ever seen by the British army.

A and B battalions had been given the objective of capturing the Village of Marcoing. The village was still largely intact and was strategically important due to its location between two main railway lines. Heavy German resistance was obviously expected, but the fighting was even more fierce than anyone had thought it would be. By the middle of the 20th November the fields around Marcoing were littered with the burning carcasses of destroyed tanks and their unfortunate crews. The first objective of the battle was to capture the 1st Hindenburg System or as it was known 'the Blue Line'. The second objective was the capture of the 2nd Hindenburg system, the Brown Line. 1st Company, 'A' Battalion had been assigned to capture the second objective. During the attack all was going well until they encountered a strongly defended German position equipped with machine guns and trench mortar. Lt. C.W. Duncan was asked to attack the position with A2 Abou-Ben-Adam II. The tank turned on the German position and with all guns firing trundled towards it. The low speed of the tank was to be it's downfall and with just 50 yards left to go the German mortars got a fix on the ponderous machine and a single direct hit put it out of action. Seven of the crew including Lt. Duncan were killed, the only survivors being one gearsman and Capt Wain who were both severely wounded. Bleeding from several wounds to the head and body Capt. Wain took one of the Lewis machine-guns from the tank and rushed the German mortar position, killing several of the enemy and taking the rest, about half the garrison prisoner. He was wounded again, but continued to fire at the retreating Germans until the Lewis gun magazine was empty. The enemy position now being cleared, he was offered medical attention for his wounds, he refused and instead took a rifle, some sources say a discarded German rifle, and continued to fire at and account for several more of the enemy. It was during this part of his remarkable action that Captain Wain was shot and fatally wounded.

The various battalion listings for gallantry medals awarded to Tank Corp soldiers after the Battle of Cambrai are truly staggering. The battle has become a legend and just about every decoration that could be won by the British Army was given out for that single battle. Officers, NCOs and other ranks won more medals and citations than seem humanly possible, but only one VC was amongst them, thus demonstrating how difficult and usually fatal it is to become a member of Great Britain's most elite military club.

The third member of the Tank Corps to gain a VC was perhaps the most famous of the four recipients, Lieutenant Cecil Harold Sewell. Cecil was the youngest of five brothers and was studying Law at London University when the war broke out. At the time he was, like many other young scholars,

Lieutenant Cecil Harold Sewell

a member of the officer training corps. By the winter of 1914, at the age of just 19, he had enlisted in the 21st Royal Fusiliers Machine Gun Section, otherwise known as the 4th Public Schools Battalion. Young Cecil did well and his military career was a whirlwind of training and preparation for the front line. He revelled in the army life and by late 1916 he had received his commission as a Lieutenant and his transfer to 'C' Battalion of the Heavy Branch of The Machine Gun Corps had been accepted. The H.M.G.C. would soon become the Tank Corps.

On the 29th April 1918 Sewell was in command of a Lincoln made Medium A Whippet by the name of Caesar II during fighting east of Favreuil. His Whippet was one of many supporting New Zealand troops attacking the German lines. During the advance Sewell noticed that a Whippet in front of his had been hit and had slipped down into a large shell crater. Perhaps in a bit of a design error, the only means of escape from a Medium A was through the rear door and this had come to rest against the side of the shell hole. Without a second thought Sewell left his own tank and sprinted across open ground under heavy shell and machine gun fire in order to help the crew of the stricken machine. On reaching the damaged Whippet, Sewell dug away at the earth around the door so that the crew could escape, all the time still under heavy enemy fire. As if this was not enough Lt. Sewell then ran back across no-mans land to where his own driver lay wounded behind Caesar II. Unfortunately, Lt. Sewell had pushed his luck too far and inevitably he was hit, but incredibly he carried on running. On reaching his driver he started to administer first aid and was in the middle of bandaging the mans wounds when he was hit a second time, this time fatally. Lt. Cecil Harold Sewell was posthumously awarded the Victoria Cross, he was just 23 years old when he died. The official citation for Sewells Victoria Cross states that *'during the whole of the period he was within full view and short range of enemy machine guns and rifle pits, and throughout, by his prompt and heroic action, he showed utter disregard for his own personal safety'*. Lt. Sewell is buried at Vaulx Hill Cemetery, near Bapaume. Both his medals and the Lincoln

Lieutenant Sewells Whippet Caesar II preserved at the Tank Museum at Bovington

made Whippet tank Caesar II are now prized exhibits in the Bovington Tank Museums collection. Caesar II can now been seen in the museums First World War Hall as a monument to Sewells selfless act of bravery. The tank is one of only five of the original 200 Medium A Whippets still known to exist today.

On the 8th August 1918 the British Army's last big push began. It was the push that would finally chase the Germans back to Berlin. The Allies had learnt a great deal over the last four years and the Germans were becoming a beaten, demoralised enemy. Once it began, as the British had hoped, the big push was enough to finally get the Germans moving in the right direction, the 8th August would later be described by General Ludendorf as *'the blackest day of the German Army'*. The battles that followed would turn the tide of the war in the Allies favour and add a final Great War Victoria Cross to the Tank Corps total.

Lieutenant Colonel Richard Annersly West VC DSO and Bar MC

Lieutenant Colonel Richard Annersly West was the fourth member of the tank corps to earn a Victoria Cross before the Armistice in November 1918. West had been born in Cheltenham in 1879 and after leaving school he joined up as a trooper with the Imperial Yeomanry. Almost immediately he was posted to South Africa for service during the Boer War. At the end of the war in 1902 he stayed in the army and continued to serve in South Africa. When the First World War started in 1914 he returned to England and gained his commission into the North Irish Horse. He was a natural leader and a fine soldier and remarkably by early 1918 West, who'd started out as a lowly trooper, was acting Lieutenant Colonel Richard Annersly West Distinguished Service Order and Bar, Military Cross with the North Irish Horse, seconded to 6th Battalion of the Tank Corps. By this time the 6th Battalion of the Tank Corps had been changed from heavy Mark IV machines to being solely equipped with Lincoln made Whippet chaser tanks.

West was recommended for a VC on the strength of two separate actions, both of which demonstrate why he had already been awarded the DSO and Bar and the MC. On 21st August 1918 at Courcelles an unseasonably dense fog descended on the battlefield. During the confusion that this caused the infantry lost their bearings. Lieutenant Colonel West was on horseback and riding through the fog he collected any men he could find and led them to their objective. During the action he had two horses shot out from under him, but continued to lead the infantry to the village, disregarding all danger from shell and machine-gun fire until the objective had been taken. As if this wasn't enough, on the 2nd September at Vaulx-Vraucourt, he again showed utter disregard for his personal safety in the face of the enemy. The plan was for West to command a battalion of Whippets into the village and beyond, exploiting and continuing the gains made by the heavy tanks and infantry on the 21st August. Prior to the tank attack he rode his horse to the infantry's forward trench to inform their commanders of the plan for the day. Upon his arrival he realised that due to heavy enemy action, officer casualties and the huge loss of other ranks there was a real danger that a German counter attack may break their lines and re-take the village. Despite the enemy being almost upon them, West rode up and down the trench in full view of the enemy and in the face of certain death, encouraging the men. He called out to them *'stick it men, show them a fight, and for gods sake put up a good fight*'. Inevitably, West was hit several times by enemy machine-gun fire and fell dead, riddled with bullets. The official citation for his VC says that '*The magnificent bravery of this very gallant officer at a critical moment inspired the infantry to redoubled efforts, and undoubtedly saved the situation'*. Although the British casualties were high, the German counter-attack at Courcelles was defeated, thanks in no small part to Lt. Col Wests inspiring example.

Lt. Col Richard Annersly West was the last Tank Corps Victoria Cross winner of the First World War. He was 39 years old when he was killed in action at Courcelles on the 21st August 1918. Less than three months later the armistice was declared and the war was over. Lt. Col West is not commemorated on any war memorial in his home town of Cheltenham.

The Tank Corps won a total of four Victoria Crosses during the years 1916 to 1918 and they are justly proud of the medals and their brethren who won them. Some may say that the Victoria Cross is a mean reward in return for a mans life. It is a small, brown, plain looking medal, no imperial grandeur, no gold, no titles and no jewels, but its looks are immaterial, the important thing about the VC is what it stands for. It is a medal that is seldom awarded and then only for the absolute ultimate in courage, for the highest bravery or more accurately for valour.

Greater love hath no man than this,
that he lay down his life for his friends.

CHAPTER VI
Bank in the Tank

War never comes cheap and in addition to the inestimable cost in human life and suffering there are also the ever-spiraling financial implications. The great loan of 1915 had raised the colossal sum of six hundred million pounds from foreign investment, but even this would only fund the war for a matter of a few months and by the end of 1917 things were getting out of hand. Great Britain was faced with a colossal national debt in excess of £6,000,000,000 with a shortfall of around three to four million pounds a day. This was largely met by appeals to large financial institutions and other wealthy investors. The Government had introduced as many measures as it dared in the form of stringent tax increases, loans from abroad and a new tax designed to kerb the excess profits made by weapon and munitions manufacturers.

In early 1916 a new body was set up specifically to meet the runaway cost of the war. The War Savings Committee realised that the global scale of the war made it a peoples war, with every man, woman and child being directly involved or at least knowing someone who was. They decided that it was time for the entire nation to rally together if we were to win and not be economically crippled afterwards. People were working longer hours and being paid much more than ever before, over one million women had entered the munitions factories and were suddenly earning wages the like of which they could never have imagined during peacetime. The War Savings Committee saw this and at the end of February 1916 they introduced War Savings Certificates and War Bonds. The War Savings Certificates were bought for the not inconsiderable sum of 15 shillings and six pence each, but could be redeemed for £1 five years later. That was only a profit of 3 shillings and sixpence over a five-year investment, but the certificates were not designed to bring profit, they were to be seen as a personal loan to the entire nation in its hour of need. Certificates could be bought from War Savings Associations that were established in schools, Post Offices and factories throughout the country. The campaign gained official symbols of a silver bullet and the Swastika, an ancient sign of good luck before it gained a more sinister roll with the Nazis 20 years later. The silver bullet and the swastika appeared everywhere, on badges, posters, paper bags and obviously the savings certificates themselves. A monthly journal was published and propaganda films produced to be taken out in mobile cinema lorries in order to boost sales. This all worked well and the money flooded in, but the War Savings Committee were always on the lookout for something new and sensational which would grab both the publics attention and their hard earned cash, the answer to their prayer would be the tank.

Ever since the first rumours of *'unstoppable armoured toads'* had been circulated in the popular press the public had been clamouring for a glimpse of the new weapon. Fanciful renditions of what a tank may or, in most cases, may not look like appeared in all of the major newspapers, Punch magazine and on patriotic postcards. The secrecy was over and the general feeling was that the time was well overdue for the new weapon to be made public. Only days after the tanks first showing at Flers in September 1916 the Daily Express had asked *'why not show one of these tanks*

Light-hearted view of the as yet unseen new 'war winner'

With no official photographs yet available cartoonists could let their imaginations run wild

Unveiled at last, a male Mk IV thrills the crowds at the Lord Mayors Show in 1917 (The Tank Museum, Bovington)

on Horse Guards Parade? Let the people see them'. The people finally got their chance to see the tank when a brace of Mk IVs took part in the Lord Mayors show of 1917. Although by 1917 the public was quite familiar with tanks on film and in the Newspapers, few people had actually seen one in the flesh. The two tanks were an absolute sensation and their curious celebrity status had been sealed.

In late November 1917, a few months after the Lord Mayors show and just a week after the jubilation's that had accompanied the tanks famous success at

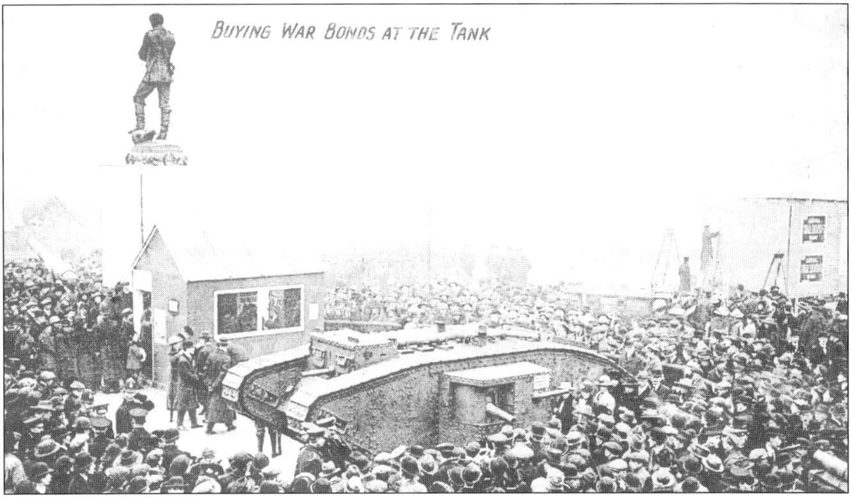

One of the Tank Banks doing a roaring trade in early 1918

the Battle of Cambrai, Trafalgar Square was the site of a War Savings fund raising event. The public could view an impressive collection of captured field guns, a German flame thrower and a machine gun salvaged from a crashed Zeppelin, but all of these curios were overshadowed by the much advertised appearance of the star of the show, the tank. The public had the chance, not only to see the tank from a distance, but to actually get close enough to touch it. Originally just one tank was present, a male Mk IV identification number 130, the machine was proposed to appear for a limited time of just two weeks before it continued on its way to the western front. However, the War Savings Committee had stumbled upon a gold mine and could not have predicted the publics fascination with the new weapon or the patriotic fervour with which they bought Savings Certificates and War Bonds. By the end of the first week the daily investment at Trafalgar Square had been over £150,000. The sponsons of the tank had been converted into cramped offices where the public could not only get their certificates stamped, but also get a close up look inside the new wonder weapon.

Somewhat inevitably, the Trafalgar Square tank became known as the 'Tank Bank' and the name stuck. The public interest in the Tank Bank and subsequent demand for certificates was so great that the War Savings Committee decided that twice as many tanks would sell twice as many certificates, so tank 130 was soon joined by a second male Mk IV. The new tank, number 113, was given a more anthropomorphic feel by being christened Julian. The tank was probably named after Sir Julian Byng who

Unlikely celebrities, Julian and crew fundraising in early 1918

Just some of the merchandise available to buy at any of the nations Tank Banks

Men with photogenic wounds line up to buy War Bonds in 1918. A variety of ranks and cap badges can be seen.

as commander of the third army, had helped to plan and execute the Battle of Cambrai in November 1917. This second machine was to become one of the First World Wars most photographed and familiar celebrities. Part of the reason for this was that instead of being a stationary attraction Julian would trundle through the crowds along previously announced routes collecting large donations from financial institutions and company offices. Julian created a phenomenal amount of War Bond investment and was responsible for not only a great deal of traffic chaos in the centre of London, but also for the untimely death of a dog who for some reason was unable to outrun the lumbering monster.

It was not only the Government who would see the financial possibilities of the tank and it was just a matter of days before souvenir manufacturers and street hawkers were selling ceramic tanks, money boxes, tea pots, hand-bags, lapel badges and picture postcards, all in Julians likeness. Even Louis Cartier was inspired by the clean industrial lines of the tank and released his now famous design icon known simply as the 'Tank Watch'. At the end of their two weeks the Trafalgar Square tank banks had raised over three and a half million pounds for the war effort. It was inevitable that after being such an unprecedented success the two tanks would continue their work and instead of the boat to France, 130 was loaded onto a train bound for Sheffield as the first stop in a nationwide tour. The Tank Banks would eventually cover most of the country selling War Bonds with tank 130 being patriotically renamed Nelson. Further tanks would soon be added to the ranks, including Iron Ration, Egbert II and Drake. One of the new Tank Banks, Egbert II, was a battered Lincoln made male Mk IV, who proved particularly popular due to scars he had received in combat at the Battle of Cambrai. The tanks are known to have been followed around the country by individual postcard and souvenir vendors, who presumably made a very good living selling tank keepsakes to the public.

Apparently Egbert II in Scunthorpe after fighting in Cambrai, but a lack of unditching beam, training number 141 and obviously drawn on machine name suggest otherwise.

So popular were the Tank Banks that the circus that accompanied them did not reach the tanks birthplace until March 1918. A letter signed simply from *'a worker'* had appeared in the Lincolnshire Echo in late 1917 imploring the powers that be *to 'include Lincoln in the tank scheme, as after all, everyone is now aware that Lincoln is the home of the tanks'.* Three months later the usually quiet streets of Lincoln came alive with a carnival atmosphere

Young and old alike flock to see the new wonder weapon (The Tank Museum, Bovington)

HMLS Lincoln sets up shop in the Cornhill at the start of War Weapons Week in March 1918 (The Tank Museum, Bovington)

during the towns 'War Bond Week' from the 4th to the 9th March. Bands played, speeches were made and acrobatic aerial displays were made by Rustons 1000th Sopwith Camel, but this was all just a sideshow to the long awaited appearance of the Tank Bank. Lincoln literally came to a standstill as thousands of people packed into the High Street to see the tank arrive in the towns Cornhill Square, just off the High Street. The usual Mk IV male was sent to collect the citizens money, but It would have been ridiculous to send Nelson or Julian all the way to a town that actually made tanks, so the Lincoln Tank Bank was simply a machine sent under it's own power from the Fosters works just a mile down the road. The tank was freshly painted and appropriately named, for the week at least, HMLS Lincoln. Unfortunately the repaint obscured the usual works production number that would have appeared on each side of the hull. By the time of Lincolns War Bond Week, Fosters were engaged solely in producing Medium A Whippets and the new Medium C Hornet, having sent the last of their Mk IV production to the tank corps in December 1917. It is thought likely that HMLS Lincoln was an old Mk IV kept by William Foster and Co as a test or demonstration machine. Throngs of locals crowded the entire route from Fosters to the Cornhill and fiscal minded property owners sold grandstand space at second story windows for a better view of the machine as it passed by. According to the Lincoln Leader newspaper '*not a single shop nor business of any kind opened its doors that day, and women and children alike joined the men in welcoming the towns distinguished visitor'*. Each town visited by the Tank Banks was given a target to reach or ideally to exceed during it's allotted week. The target differed according to the size or wealth of a town, but for a medium sized, reasonably affluent town, it was often the price of a new Destroyer, which would cost £150,000 to build. The people of Lincoln had been charged with raising this amount and they were not to disappoint, as by the end of its first day in the Cornhill HMLS Lincoln had taken a staggering £45,903. On the second day the tank was to prove even more of an attraction as, preceded by its military guard and the Ruston Prize Band, it moved off towards the Cattle Market on Monks Road. It was to have begun its journey at 11.15am, but so great was the public interest that by 10.30am the police had decided that it would be dangerous to keep the crowds waiting any longer and the proceeding began. Various dignitaries including the Mayor of Lincoln and William Tritton followed behind by car, whilst three Sea Scouts and several Tank Corps soldiers rode proudly on top of HMLS Lincoln. Once again the City stopped as the lumbering machine travelled up the High Street, almost the entire population of Lincoln and thousands of visitors from elsewhere lined the tanks route. It reached the top of the High Street and turned right

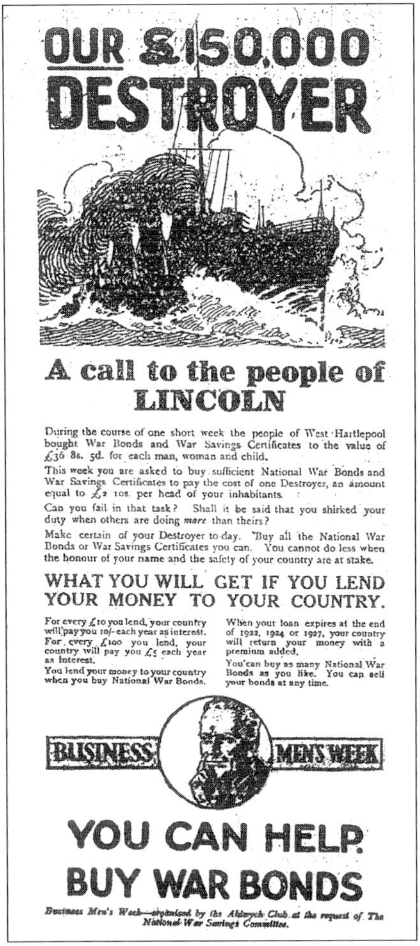

at the Anglo-American flag adorned Stone Bow, onto Saltergate, then up the appropriately named Bank Street onto Silver Street, all the while the band played and the crowds cheered. For the last leg of the journey the spectacle of the tank was joined by an aerobatic display from Rustons 1000th aeroplane. Like many people at the time Joseph Ruston was a keen Egyptologist and the aeroplane had been painted to reflect this with the Behudet, a winged sun, emblazoned across its wings. The Ruston Sopwith Camel dropped leaflets urging people to buy more war bonds, as the tank took its place in front of the main auctioneer's stand at the Monks Road Cattle Market. HMLS Lincoln had taken almost an hour to travel just under 1 mile. Tuesday was market day and a ready-made audience of farmers and land-owners were packed into the Cattle Market. The usual round of speeches now followed with the Mayor starting the proceedings, after which came Mr. W. A. Richardson and Mr. J. E. Walter, the Cattle Markets head auctioneers and two of Lincolns best known residents. The two auctioneers spoke briefly about the John Bull spirit, the alleged German atrocities in Belgium and the need for businessmen and labourers to give generously in this, the Country and the civilized Worlds hour of need. After their turn on top of the tank Mr. Walter put his words into practice and bought £500 worth of certificates.

There is no doubt that throughout Lincolns War Bond Week the tank was the main focus of everyone's attention and the public excitement that accompanied it grew steadily over the days to come. However, Lincoln reserved its most enthusiastic welcome for its adopted son and the co-creator of the beast,

Lincolns Tank Bank goes mobile on its way to Monks Road (The Tank Museum, Bovington)

The High Street has changed little since Lincolns long awaited visitor made its ponderous way north towards the Stone Bow (The Tank Museum, Bovington)

Sir William Tritton. Only a few months before, in acknowledgement of his newly conferred knighthood, Tritton had asked his co-directors at Fosters to allocate a sum of money from the next balance sheet for the endowment of a bed in Lincoln Hospital. The bed was established with an endowment of £500 and Tritton had received a rapturous welcome from public, patients and hospital staff alike as he had stood up to make his dedication speech

Rustons workforce with their thousandth creation, complete with Egyptian colour scheme

The Tank Bank continues its slow march to the Monks Road Cattle Market (The Tank Museum, Bovington)

and unveil a brass plaque marking the installation of the Fosters Tank Bed, the plaque that Tritton unveiled still hangs on the wall of the Hospitals main entrance corridor. The response from the crowd at Lincoln Hospital would be nothing compared to the welcome Tritton received at the cattle market

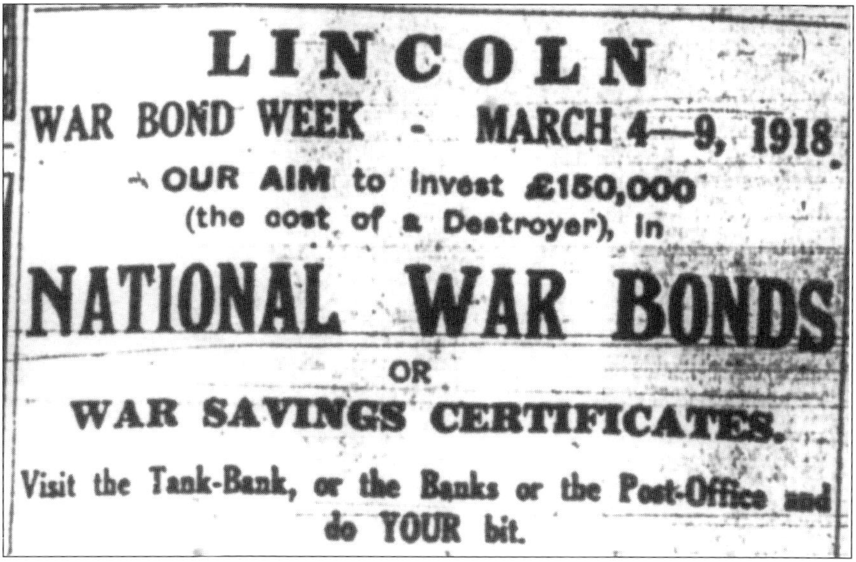

LINCOLN

WAR BOND WEEK - MARCH 4—9, 1918

OUR AIM to invest £150,000
(the cost of a Destroyer), in

NATIONAL WAR BONDS

OR

WAR SAVINGS CERTIFICATES.

Visit the Tank-Bank, or the Banks or the Post-Office and do YOUR bit.

and after Mr. Richardson and Mr. Walter had said their piece they introduced Tritton to the crowd. The applause and cheering went on for several minutes before order was regained. Tritton started off by removing his coat, *'I am here to appeal to your hearts, your pockets and your patriotism. I have taken off my coat to emphasize to you that I am serious and that you must be serious in standing up for your county and your country. There are three classes of people in the country now, there is the fighting man, the munitions worker and those standing before me. The man at the front and the munitions workers are doing their bit and this is the chance for the normal man and woman in the street to do theirs'.* Anyone who ever questioned that Tritton was a forthright man who said what he liked and liked what he said, would be left in no doubt by the conclusion to his rousing speech. He pushed the people of Lincoln further by reminding them that the Destroyer that they were raising money for was especially important to them as they lived *'just 30 miles from the North Sea and without the men of the Royal Navy and their Destroyers, what was there to stop the German hoards from ravishing your villages? Without these men, your mothers and children will suffer bloody deaths and your wives will stand a great probability of being dishonoured'.* Trittons scare-mongering speech was a tremendous success and the applause as he clambered down from the top of HMLS Lincoln was deafening. The speeches had been well received and queues of citizens kept a steady march past the tanks open sponsons every day that week. Specially created souvenirs were available including Lincoln in a tank, various views

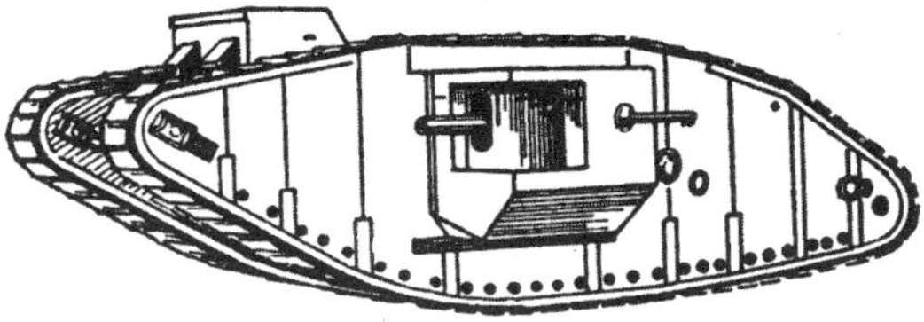

Tank Anthem.

THE Glorious Foster Tanks
Have smashed the German ranks ;
These fighting Tanks,
Their noses to the Rhine,
Have stormed the " Pill Box " line,
And done their work so fine,
The wondrous Tanks !

Thousands of lives they save ;
These Tank Crews all so brave,
Great deeds perform.
Through barb'd wire, shot and shell ;
Huns thought impregnable,
The Landships fought right well :
Let's cheer these Tanks.

Our Tanks are " doing their bit,"
Making a famous hit
In this Great War.
When this " big job " is done
We'll shout " Bravo ! Lincoln "—
" Something made to smash the Hun,"
Which England thanks. G.T.

PRICE 1d.

The profits from the sale of these Cards are devoted to the Comforts Fund inaugurated by William Foster & Co., Ltd., Lincoln, for the benefit of Soldiers, Sailors and Tank Boys at the Front.

The William Foster Tank Anthem postcard

of Lincoln that unfolded from a picture postcard, and the Tank Anthem. The Tank Anthem was a patriotic poem read to the tune of the National Anthem and had been specially written and printed by Fosters on postcards that were sold for one penny each. The money raised from the sale of these postcards went towards the comforts fund set up by William Fosters for the benefit of Soldiers, Sailors and specifically, tank crews. It is not certain who penned the Tank Anthem as it is simply signed G.T., but a second illustrated poem released by Fosters was titled 'The Centipede' and it shows a surprisingly playful and creative side to it's author. The initials at the end of the piece are W.A.T. clearly standing for William Ashby Tritton.

A centipede was happy quite until a frog in fun said 'Prey which leg comes after which?'
This raised her mind to such a pitch she lay distracted in a ditch considering how to run.
The Tank, another Centipede was certain how to run.
He understood that right was might, without ado he meant to fight,
Spontaneous he went on sight, bald headed for the Hun!

Another fund raising method that was possibly unique to Lincoln was the production of small bronze and brass souvenir tanks. These were produced by the workers at Fosters from moulds supplied by the company. Old bronze bushes or scraps of brass left over from tank manufacture were melted down and poured into the moulds to produce six-inch (150mm) long models of a male Mk I tank complete with steering gear. These models were sold for 5 shillings each and eventually lead to the employees of Ruston and Proctors

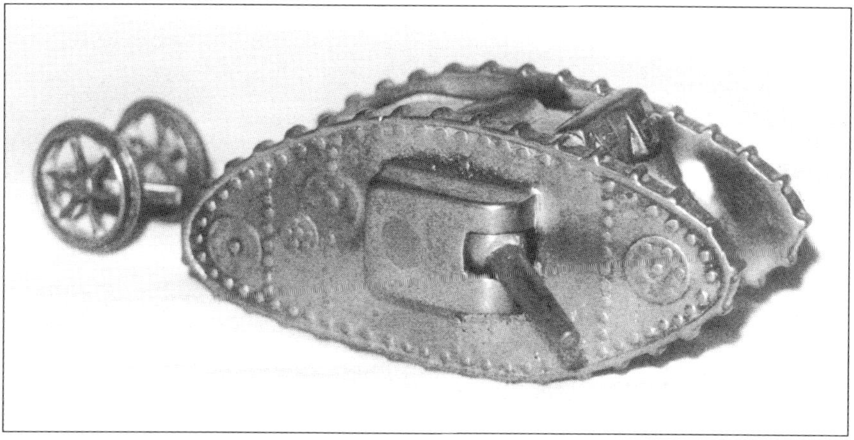

Fosters mini-tank. Made from old scrap copper bushes and in 1918 could have been yours for just 5 bob

producing bronze and aluminium models of their main product, Sopwith biplanes. The models of both the tanks and the aeroplanes sold well during Lincolns fund raising week although surviving examples are very rare today. For those with a little more to spend than 5 shillings there were Savings Certificates and War Bonds. Private investors bought anything from the standard 15s 6d certificate to one unnamed woman who bought £820 worth

and paid in gold Sovereigns. Mr. Richardson and Mr. Walters appeal to Lincolns businessmen had not fallen on deaf ears either as car and motorcycle dealer R.M. Wrights and Co. LTD pledged to give every penny spent at their dealership during the entire War Bond Week to '*Swell the Tank Fund!*'. The publicity that surrounded the tank in Lincoln was unprecedented, but The Lincolnshire Echo newspaper did one better than most and actually managed to get an interview with the tank itself. In a curious piece entitled *Lincolns Cornhill tank; An interviewer meets his match,* HMLS Lincoln is quoted as saying that *'this is the town of the tanks, the place from which I came should make an overwhelming success of War Bonds Week. He who makes two blades of grass grow where only one grew before is doing a mighty useful service, and the city that is asked for the price of one Destroyer and responds with two is a city that an old veteran like me will be proud to belong to'.* The people of Lincoln must have heeded the tanks wise words as by the end of War Bonds Week they had invested enough money to buy not two, but three Destroyers, and by the end of the war they had raised over £5,000,000. One of the thousands of patriotic Lincoln residents who did their bit during War Bonds Week was Bill Baker. Bill, who only recently passed away aged 91, would recollect that when he was five he and his mother visited the Cornhill Tank Bank. He always remembered climbing into the tanks interior where he paid his 15/6 to buy a Savings Certificate. Young Bill made good use of his investment and in 1922 cashed the Savings Certificate in at his local Post Office for £1, which he promptly spent on a pair of new football boots. Bill could vividly remember his first hand encounter with the Tank Bank and its enormous impact on Lincolns War Bonds Week in 1918. A few days after buying the Savings Certificate Bill was leaving St. Andrews Infants School on St. Andrews Street in Lincoln when he spotted the tank turning off Canwick Road into Portland Street on it's triumphant return to Fosters after a successful week raising money for the war effort. He became part of a large band of cheering schoolboys who escorted the ponderous machine all the way back to its Firth Road home. This short trip of less than a mile must have taken some considerable time as Bill suddenly realized to his horror how late it had become. Luckily when he returned home his Mother was so relieved to see him safe and sound that he escaped any punishment.

CHAPTER VII

A tank in your Town

After almost four gruelling years, hostilities finally ended on the 11th November 1918. The Tank Corps had won 4 Victoria Crosses and lost hundreds of brave men. Many, including the defeated enemy, said that the tank was the weapon that had won the war for the allies. Monsieur Loucheur, the French Minister of Munitions wrote that '*There are two kinds of infantry. Men who have gone into action with the tanks, and men who have not. The former never want to go into action without tanks again'*. To claim that the tank won the war is a very simplistic view when there are so many other factors to take into account. The subject still sparks heated debate even today, but no single weapon can win a war, the tank was merely a weapon that helped to open up the war and got it moving again. Despite the controversy one thing is quite beyond question, the fact that the tank was a truly extraordinary invention. On the right ground, with the right crew and well considered artillery and infantry support it could be a devastating weapon. A second, perhaps even more impressive, face of the tank phenomenon was its effect on the home front. As a moral booster or propaganda tool the tank had been invaluable and the publics fascination with it had raised millions of pounds for the war effort.

Newark's female Mk IV souvenir tank triumphantly enters the Castle grounds in 1919

A sad sight as Newark's female Mk IV tank is reduced to scrap after less than 20 years on display (Newark Museums. Newark and Sherwood District Council)

Unfortunately, despite the public fascination for the tank there were those in Government and in the Military who could no longer see the need for the Tank Corps or its tanks now that victory had been won. Suddenly hundreds of training and fighting tanks were made redundant and rusting hulks littered Salisbury Plain, Bovington Camp and the training camps in France. The war department found an answer to the disposal of the tanks by giving them away to towns and cities that had raised significant amounts for the War Bonds campaigns. During the Feed the Guns campaign in October 1918 the idea had been put forward for the famous tank bank Egbert II to be given away as first prize to the city

The people of West Hartlepool claim their prize as Egbert II is presented to the town in 1919

or town that raised the largest amount per head. In the event General Ernest Swinton presented the Lincoln built tank to West Hartlepool where the citizens had raised the remarkable sum of £31.0s.1d per head. The souvenir tanks were finally distributed, but were not popular with everyone and it was said by many that they were neither memorials nor war trophies. By the early 1930s the tanks were beginning to corrode and it was generally considered that they did not warrant the constant expense of their upkeep, the female Mk IV Cambrai veteran given to Newark in 1919 was a typical case. Newarks tank had taken pride of place in the castle grounds and managed to survive until 1937 when a letter appeared in the Newark Advertiser saying that *'the rusting hulk is obscuring the view of the park'*. Within two months Lees Scrap of Grantham had cut it up and taken it away, at a cost of £33 to Newark Council. For those tanks lucky enough to survive a similar fate, it would only be a stay of execution as they were systematically destroyed during the scrap metal drives of World War II. Of the 265 Tanks distributed

On a cold February morning in 1931 the tank that had stood behind the British Museum in London since 1919 is uncermoniously sliced up for scrap

on behalf of the War Savings Committee only one still stands in its original position today. The Mk IV female given to Ashford in Kent was placed just off the High Street where it was soon adapted to house a small electricity sub-station. As it was performing a useful function it was spared during the Second World War and has recently been refurbished. The Souvenir Tanks

Guarding Ashford High Street, the last of the souvenir tanks still standing in its original position

were almost exclusively females as they were the most common variant and their guns could be easily removed without detracting from the overall look of the machine. Communism and Red Revolution were a very real problem in the days just after the Great War, the Russian Revolution of 1917 was still fresh in people's minds and there was a worry that deactivated 6-pound guns could quite easily be put back into service and used by undesirables should the same thing happen here. Several towns did receive the more warlike male tanks, but not without good reason. Coventry made the Daimler engines and Newcastle made several marks of tanks, both of these towns received male machines. Lincoln also qualified as one of only a handful of places chosen to receive a male as the idea of the tank had first been realised and the original machines had been created in the town. A Souvenir Tank was specially requested by the parishioners of Boultham in Lincoln, during the war Fosters had bought several acres of land, including a quantity of the nearby Boultham Park, and transformed a portion of it into an extension to their machine testing ground, so the plans for a memorial in the park were thought appropriate by many. The ambitious idea was to create a peaceful, yet striking oasis formed by a semi-circular recess, backed by a stone wall, fronted by a semi-circle of trees with seats running along its length. In the centre the plans specifically called for '*a well worn, war used tank to be displayed on a stone pedestal flanked on either side by captured German Howitzers*'. A bronze memorial plaque bearing the names of those from the area who had fallen in battle was to be fixed to the side of the tank. However, spiralling expense and difficulty in procuring the weaponry meant that after some considerable debate the Boultham plans came to nothing. Perhaps if the residents of Boultham had succeeded in their plans to create an official war memorial with a tank at its centre it would have escaped the scrap-man and still remain today. There are stories that Lincoln actually received two presentation tanks with the second elusive tank being placed on the Abbey playing fields off Monks Road although photographs of the area taken at the time show only captured German Howitzers with no sign of a tank.

Lincoln's one and only souvenir tank finally arrived in great secrecy several days before the dedication ceremony on Wednesday October 28th 1919. The Lincoln machine arrived by rail from Bovington and was taken by traction engine and trailer to the Arboretum on Monks Road where it was kept until being towed to Wickham Gardens off the historic Bailgate. While at the Arboretum the sponsons were removed as a Mk I tank including sponsons would never have fitted through the gate at Wickham Gardens. Joan Barratt was almost 4 years old at the time and remembers seeing the machine arrive at the Arboretum. The difference in opinion between monument and killing machine was rife in the crowd that day and Joans mother hurried her passed the tank saying crossly *'I don't know why that thing's here, we've had enough of war'.* Joans mother had hit on the main problem with presenting tanks to towns. The machines were seen by many as memorials or in this case an example of Lincolns proud engineering heritage. However, the gifts were understandably seen by others as cruel and hellish reminders of four years of bloody strife and death. Many veterans and their families objected to the light-hearted view of the tanks as monuments and some even planned their own more discrete commemorations. In 1918 the parents of 200988 Private Leslie Wray had a stained glass window installed in their local village church at Nocton, about 6 miles south of Lincoln. On the 20th November 1917, the first day of the Battle of Cambrai, Wray was the driver of female Mk IV tank E27 'Ella'. Private Wray was to become one of the first casualties of the battle when at 6.20am, before the tank had even begun to move off from the starting point, Ella received a direct hit to the front from a German heavy field gun. This one shot injured six of the eight man crew and instantly killed Pte. Wray and the officer commanding the tank, 2nd Lieutenant W.S Haining. Touching personal memorials such as Pte. Wrays stained glass window seemed more fitting to many people than a dirty great tank placed in the high street, therefore the positioning of the presentation tank was crucial. For this reason the tanks were usually carefully placed off the main road in a quiet part of the town, but not so well hidden that nobody knew it was there at all, public parks were a favourite choice. After much deliberation, Wickham Gardens, off the historic Bailgate, had been chosen as the site for Lincolns souvenir tank. The Lincoln Leader newspaper for November 1st 1919 reported that when the tank arrived at Wickham Gardens it was well received by the citizens of Lincoln, and that the *'over the top Mk I veteran of the Somme had been specially chosen as it was one of Lincolns own creations and had finally returned to its birthplace'*. A crew of six men from the Tank Corps conducted the delivery and decommissioning of the tank. The operation was commanded by Capt. W. F. Farrar M.C. assisted by Lieut. Meake, a Sergeant and three mechanics. Captain Farrar must have been quite a Tank Corps celebrity as he was the commander of Mk IV F.22 'Flying Fox' at the battle of Cambrai. The ill-fated tank was one of twelve machines ordered to take the road bridge at Masnieres on the 20th November 1917. Unfortunately, damage already sustained during the battle and the considerable weight of the tank caused the bridge to collapse whilst F.22 was still crossing. Farrar managed to evacuate his crew and fought his way back to the British lines without losing a single man. Flying Fox and the collapsed bridge were not immediately removed by the Germans who instead used them as part of the foundations for a new bridge.

Flying Fox at the start of it's new career as a bridge support after Cambrai in 1917

As he gave his official hand over speech and offered the battle-scared old tank to the people of Lincoln Capt. Farrar must have been all too aware of the mixed emotions raised in the assembled crowd. Farrar seemed to be doing his

his best to justify the tank as a monument when he said, *'People must not forget the great opposition there has been throughout the war to inventions, not only in tanks, but aeroplanes and everything else. In the tank, unlike some of the weapons the Boche used, we had a clean weapon of warfare. There is nothing about it to remind us of poison gas or liquid fire and criticism of the tank is unfair, to the men who went over the top in France, the men who fought in the tanks and to the brains who had invented the tank. People say they do not want to be reminded of the carnage and cruelty of war. There is much to be said for that point of view, but you must not forget that, outside stands this old tank, and it will always stand there, as an ever-lasting symbol of British industry, resource and pluck. It stands on the green, as it will throughout the years and centuries to come, it will always be a symbol from 'out there' of what had been done by the people who had stayed at home'.* This ever-lasting symbol did not stand for centuries, it stood in Wickham Gardens for as little as six years. There are stories that the tank was still in place during Lincoln's Empire Day celebrations on May 24th 1935, but some say that by the mid 1920s it had begun to look quite shabby and it was in fact scrapped in mid 1926? The steering gear at the back of the machine was apparently a natural ladder leading to the top of the old tank, consequently a string of injured children and spiralling upkeep combined to seal its fate and Lincolns tank went the same way as most others, being unceremoniously scrapped. The handover ceremony of Lincolns souvenir tank took place on a day of driving rain so no photographs were taken at the time and despite standing on show for several years, the Lincoln tank appears to have been the only souvenir machine not to have been the subject of a single photograph. Despite exhaustive research, radio and newspaper appeals by the author, the Lincoln Tank Group and the newly formed Friends of the Lincoln Tank no photograph showing the Lincoln tank at Wickham Gardens has yet been found.

CHAPTER VIII

Where are they now?

The first generation of tanks has not survived well, they were made for a specific purpose and they have obviously had a very rough working life. They were hurriedly and consequently, often poorly constructed making prolonged maintenance a virtually constant and prohibitively expensive exercise. Many parties blamed the inexperienced workers in the tank factories and William Tritton would be forced to defend Fosters quality standards. He admitted that items such as the track rollers may well have been under engineered, but this was due entirely to the fact that the original War Office specifications had stated that tanks were to have an operational life span of no more than 50 miles (80km) maximum. However, it was not uncommon for training tanks to cover twice that mileage in just a single week. Albert Stern likened the first tanks to a missile, '*once they were used that was it, they should not be reused, they should be immediately replaced*'.

With the eventual end of the Great War in 1918 hundreds of tanks of all marks suddenly found themselves out of a job. Many were given away to towns as souvenirs and many found their way onto military firing ranges as target practice for the next generation of machines. Most were simply scrapped where they stood, gangs of five or so men being employed to break up each tank by hand with fourteen-pound sledgehammers. On average, it would take about a week to break each rivet and reduce a tank to its component parts. However, as the gangs were usually employed on piece work the record for disposal of a single tank was an astounding two and a half days. A clue to the amount of old machines that must have littered the fields around Bovington Camp is that despite the hurry to dispose of the old machines the gangs work was not over until fourteen years after the end of the war, the last tank finally being broken in 1932.

Some old tanks managed to avoid the scrap man and find themselves useful, if temporary, post-war employment. One Mk IV was used for some time as a towing tractor for airships at the Royal Navy Air Station at Pulham in Norfolk. A brace of Mk IV females were purchased as War Surplus by a consortium of ex-Tank Corps officers. The two tanks were fitted with wooden top-decks and spent a curious few years giving rides to fare paying passengers along the promenade at Southend-on-Sea. A battered old Mk I, minus it's steering gear, was purchased by a building contractor in Worcester. He used the machine for demolition work, apparently simply charging straight at the building in question until it collapsed. He and his 'demolition tank' became local celebrities until one day in 1919 when he charged into a large detached property in the middle of Worcester and disappeared through the floorboards into the cellar. He escaped through the rooftop hatch, but the tank was wedged fast and quite irretrievable. After several unsuccessful attempts to extricate the stricken machine the cellar, including it's new occupant, was filled with concrete and new houses built over the top.

In 1940 Sir Donald Bailey invented his famous Bailey Bridge, a brilliant system of interlocking bridge sections supported on floating pontoons that could, with a little training, get troops and vehicles across otherwise uncrossable rivers in a matter of hours. In order to test the weight bearing capabilities of his new invention something heavy, but utterly worthless was required to be driven across it. An old Mk V tank was loaded with pig iron and the controls jammed open. The Bailey Bridge did not collapse and passed the test with flying colours, the ultimate fate of the unfortunate Mk V is not

Fresh from the battlefield, a Medium Mk A Whippet and a male MkIV at the Musee de Royale de l'armee, Brussels

recorded. This kind of treatment was typical of the way the old tanks were viewed by the military. Like many of the men that had fought alongside them, they returned home to an indifferent welcome and their achievements were soon to be forgotten. Even today, the tanks position is far from comfortable and it is still having to fight for its survival. Senior Army officials recently stated that large heavily armoured battle tanks are hardly required in modern warfare and they will eventually be phased out in favour of small, high speed chaser type machines using ultra light carbon-fiber armour. Some in the military have gone further still by predicting that terrestrial fighting vehicles will soon be scrapped altogether to make way for swarms of attack helicopters. The tanks entire reason for existing was its ability to cross broken ground and take to battle to the enemies front door, but modern inter-continental missiles have made the tanks main ability redundant. Even the latest technology in tank design could soon be consigned to museums to stand alongside its illustrious forefathers from over 90 years ago.

Sgt. Harry Emans began in the Motor Machine Gun Corps in January 1916 and fought in every major tank battle of the Great War. Here he is seen in Cologne 1919 waiting for a train to take him back to Blighty and an uncertain future (Ray Hooley)

Probably the best collection of tanks in the world and by far the best collection of surviving First World War British tanks can be seen at the Tank Museum in Bovington, Dorset. Many other examples do exist scattered around the world, from Ashford in Kent, to Canberra in Australia. More probably still exist still buried on the battlefields or hidden away in scrap yards, but the following list covers all variants of surviving principal heavy and medium British First World War tanks known to the author at the time of writing.

Model or Mark	Variant	Number known to still exist	Present Location
Little Willie	N/A	1	Bovington Tank Museum, Dorset
Mk I	Male	1+	Bovington Tank Museum, Dorset
	Male		(Fragments) J. Letailles private museum, Bullecourt, France
Mk II	Female	1	Bovington Tank Museum, Dorset
Mk IV	Male	7	Bovington Tank Museum, Dorset
	Male		Musee de Royale de líarmee, Brussels
	Female		Australian War Memorial, Canberra
	Female		Museum of Lincolnshire Life, Lincoln
	Female		Ashford City Centre, Kent
	Female		Flesquieres, France
	Female		The Aberdeen Proving Ground, USA
Mk V	Male	8	Bovington Tank Museum, Dorset
	Male		Imperial War Museum, London
	Male		Regional History Museum, Ukraine
	Female		Regional History Museum, Ukraine
	Female		Archangelísk, Russia
	Hermaphrodite		Kubinka Tank Institute, Russia
	Hermaphrodite		Two MK V Tanks are displayed in the
	Hermaphrodite		city of Lugansk, Russia
Mk V *	Male	1	Patton Museum of Cavalry and Armour, USA
Mk V**	Female	1	Bovington Tank Museum, Dorset
Mk VIII	N/A	2	Bovington Tank Museum, Dorset
			Aberdeen Proving Ground, USA
Mk IX	Supply	1	Bovington Tank Museum, Dorset
Medium A Whippet	N/A	5	Bovington Tank Museum, Dorset
			The South African Military Collage
			The Aberdeen Proving Ground, USA
			Musee de Royale de líarmee, Brussels
			Base Borden Military Museum, Canada

Probably largely due to the number produced, the Mk IV and V tanks appear to have had the best survival rate. Considering that William Foster and Co Ltd only made 100 Mk IVs it would seem improbable that any Lincoln built machines have survived the ravages of war and the intervening years, but at least one does still exist. The Mk IV male currently on display at the Tank Museum at Bovington bears the identification number 2354 showing that it

The Lincoln made HMLS Excellent at the Tank Museum, Bovington

was built in Lincoln in August 1917. This particular machine is thought to have spent 1917-18 as a training tank at Bovington Camp and after the war it was presented to the Royal Navy Gunnery School HMS Excellent at Whale Island, Portsmouth in recognition of the camps help in training tank crews to fire 6 Pdr guns. Tank 2354 was christened HMLS Excellent and stood for many years as the camps gate guard. In 1940 Lieutenant Menhinick, an officer at Whale Island, hurriedly pressed the tank back into service for a second war by scavenging parts from another tank on Southsea Common. The tank was reputedly used for the defence of Portsmouth, but the Mk IV was never really designed for use in confined spaces and after some unfortunate incidents involving the demolition of a pair of school gates and the tank running into someone's car, the machine was returned to relatively harmless patrol duties in the wide open spaces of Whale Island. After the Second World War the tank stayed at Whale Island until finally being donated to the Bovington Tank Museum in 1974. The museum gave HMLS Excellent a complete overhaul to working order and in the mid-1980s the tank came out of retirement for a starring role in the BBC television series Soldiers. Tank 2354 is currently painted to depict its days as a training machine and instead of a name carries the correct three figure training number 102 on either side.

Perhaps one of the most interesting of the other surviving Mk IV tanks is the example on display at her spiritual home of Lincoln. The Cambrai veteran

Flirt II in training before the battle of Cambrai in late 1917 (The Tank Museum, Bovington)

F4 Flirt II was a replacement tank for her original namesake, Flirt, which had been ditched and abandoned near Ypres in June 1917. On the 24th November 1917 Flirt II was one of 17 tanks sent to attack Bourlon a few miles west of Cambrai. She and her crew had already seen action in the early days of the Battle of Cambrai, but the 24th November was not to be a good day for the eight men in Flirt II. As the tanks and their infantry support approached the village one of Flirts fellow tanks, F13 Falcon II, was forced to turn in order to avoid a wounded soldier lying in the road and slid sideways down a hill, finally coming to rest wedged against a low wall. The tank was unable to gain any traction and found itself sliding helplessly around in the mud. The crew attached the unditching beam and after a couple of revolutions the chains on one side of the beam snapped, things were not going well and they were set to become much worse. Flirt II came to Falcons rescue and attempted to tow its cohort clear. The ground must have been much worse than anyone had realised as Flirt also became stuck, after some time the crew managed to free her, but the strain of extricating herself and towing the other tank was too much for Flirts transmission and the gears became stripped. A third tank, F1 Firespite II, now joined the group with their efforts descending into farce. However, the situation was far from amusing and things were not helped by the fact that the infantry that had been supporting them was now ordered to withdraw leaving the three tanks and their crews struggling in plain view and becoming the target for every German machine gun in the area. Suddenly Firespite developed engine trouble and there was serious danger that the three tanks would be lost in the first moments of the battle without even firing a shot. Firespites commander decided that discretion was the better part of valour and managed to limp his machine back to British lines to fight again another day. Meanwhile Flirt and Falcon were stuck and under heavy fire, the crews turned the tanks into temporary pillboxes and bravely kept up the fight for almost an hour before they were forced to retreat due to low ammunition and sheer force of enemy numbers. After the battle Flirt II was retrieved by a British recovery crew making her one of only 18 of the original 42 F battalion tanks to survive. Falcon II was less lucky and was captured by the Germans and taken to their workshops at Charleroi in Belgium. Before being repainted and becoming a Beutepanzer, Falcon II briefly became a film star, ironically appearing in a propaganda film promoting German war bonds. After her recovery Flirt II was fully repaired by the Tank Corps Central Workshops and returned to her crew for further action. In early 1918 Flirts battalion was renamed 6th instead of F battalion and was reassigned to Medium A Whippets instead of heavy tanks, for this reason Flirt II and the remaining MK IVs of her battalion were eventually disbanded and returned to England, Flirt II was to see her final action in March 1918 at Arras. After

Flirt II, Tank Corps Officers and Mascot prior to Cambrai

the war Flirt II was one of the unfortunate old machines that was sent to Lulworth Firing Range to be used for gunnery practice. By 1932 Flirt II had not been obliterated by trainee gunners and therefore she was reprieved and returned to Bovington Camp 11 miles (18km) down the road. She was towed back to the camp by a Gainsborough made Mk IX supply tank, between them they took 11 days to complete the short journey. The next chapter in Flirt IIs story was to be a more restful one as for many years she stood as gate guard at the Tank Museum at Bovington. In 1983 after talks between the Lincoln Tank Group, Lincoln City Council and the Bovington Tank Museum it was agreed that Flirt II should be given to the City of Lincoln on permanent loan. Once she arrived at Lincoln she was transferred to the workshops of Ruston Gas Turbines where the apprentices spent almost two years restoring her. The original plan had been to restore the tank to running condition, but the engine and gearbox had been removed at some time in the past and the restorers were unable to find replacements, so the machine remained a static exhibit. After a brief, but successful appearance as one of the 'Treasures of the Nation' at the British Museum in London, the tank was returned to Lincoln and eventually housed in the Museum of Lincolnshire Life on Burton Road in Lincoln, where she is one of the museums most prized exhibits. Flirt II is a female tank and subsequently was not built in Lincoln; she is most likely to have been built by the Metropolitan Carriage, Wagon and Finance Company of Birmingham. However, she now stands as a fitting tribute to the ordinary people from Lincoln and all over the United Kingdom who built her and hundreds of others like her.

Flirt II during her restoration at Rustons in the 1980s

Flirt II, like most of the tanks that still exist, is a remarkable survivor, but perhaps one tank stands out above all the others as a truly exceptional story of survival, detective work and perseverance. During the exhaustive research for their excellent book *'Following the Tanks, Cambrai'* Philippe Gorczynski and Jean-Luc Gibot spoke to several survivors of the war with vivid memories

Back in her spiritual home of Lincoln, Flirt II now helps to teach a new generation about the sacrifices made during the Great War (Museum of Lincolnshire Life and The Tank Museum, Bovington)

of the Battle of Cambrai. They became increasingly convinced that although the British and German recovery crews had retrieved most of the salvageable tanks knocked out during the fighting, many of the badly damaged ones remained where they had fallen and were eventually buried. One particular story seemed just too good to ignore. A Madame Marthe Bouleux remembered that during the war when she was 17 she had watched a tank being pushed into an enormous hole in a field near the village of Flesquieres. The village had been the site of bitter fighting in 1917 with D battalion losing 24 of their 35 tanks on the first day of the battle. Philippe and Jean-Luc used infrared film to photograph the area and made metal detector surveys of the individual fields. They kept coming to the same conclusion; there was a very large metallic object in the middle of one of the fields to the south west of the village. An excavation was planned involving local heritage and archaeological groups and on the 5th November 1998 they succeeded in locating a World War One tank still buried on the battlefield after 85 years. The race was now on to uncover the rest of the tank before the 20th November, which would be the 81st anniversary of the battle in which the tank had become a casualty. The tank was fully excavated in time and a touching ceremony took place on the evening of the 20th during which a wreath was laid in memory of the soldiers who had died serving with the Tank Corps during the Great War. After extensive research and some fortuitous good luck Philippe and David

Memorial to the fallen of the Tank Corps, Deborah on display in Flesquieres (Association Tank de Flesquieres, France)

Fletcher from the Tank Museum at Bovington managed to identify the tank as D51 Deborah. The tank was carefully lifted and is now part of a growing museum in a barn in Flesquieres only a few hundred yards from where the tank was found. Deborah can now be seen by arrangement with Philippe, but she is not to be restored, rather conserved in *as found* condition. The damage to her twisted hull goes to demonstrate the destructive power of the German guns and the relative frailty of these early tanks. She is a fitting monument to the brave Tank Corps soldiers who fought in France and also to the people who created her in England. Other tanks are known to still exist below the fields of France and Belgium, perhaps one day someone will copy Philippe's example and excavate them? However, this is an undertaking that should not be taken lightly, even after funding has been raised and a tank has been located, many other factors must be considered before any excavation can take place. The tanks are very often war graves, many still containing the remains of their crew. The proper care, possible identification and reburial of human remains must be at the fore of any archaeological investigation. In addition to the possibility of human remains, serious thought must be given to the safe disposal of any live ordinance that will undoubtedly still be on board. Every country has different laws concerning the excavation, display and transportation of archaeological artifacts and despite the fact that you located and excavated it, legal ownership of an exhumed tank may be difficult to establish. You may end up spending many thousands of pounds and many hundreds of man-hours on a tank that cannot be exported out of the country it was recovered from. Even after the excavation is done, the ownership has been established and the tank has been successfully exported the long term care and display of such a large item will be a major problem. Whilst underground the metal is usually in anaerobic conditions and the rate of corrosion is often stabilized, but once the now fragile 90 year old metal work is exposed to the air it will start to corrode at an accelerated pace. Despite the fact that most of us would dearly love to see more Great War tanks on display around the world perhaps it would be better if we were to let them lie where they are now, undisturbed and preserved in-situ below the ground for future generations lest we forget?

Deborah is a rusting, shell torn reminder of the horror of the trenches and by no means a typical pristine museum piece, but after the hard working life they have had very few tanks are likely to be in the same condition as they were in when they left the factory. The Mk IV and the Mk V at the Bovington Tank Museum are both still in running order, although sadly they are unlikely ever to be used again due to the obvious strain on their superstructure and the wear sustained by their engines, rollers and tracks. The fact that the machines are still capable of running on their original tracks at all is a testament to the design and engineering prowess of the man who created the system over 90 years ago.

The two Mk Vs displayed in the city of Lugansk in Russia are the remains of a number of tanks given to the Russian Tank Corps during the First World War. Brand new Mk Vs, Whippets and several Medium B machines were given to the White Russians to help in their struggle against the communists, with a significant number having an operational career well beyond their projected lifespan. They paraded across Red Square in 1928 and many examples of the Mk V were widely used in sub-zero conditions right up until the late 1930s, some even apparently being fielded against the Nazis during the defence of Stalingrad in 1942. The Russians called these tanks 'Rikardos' as the only identification they could find on them was the Ricardo trade-mark cast onto the engine rocker box covers. Not only were many old tanks employed during the Second World War, even the old design team was called back together. Fosters had kept their hand in by designing a tank for the Italian army in 1937, with William Rigby heading the design team for the project, Rigby tested the tank on the Grand Cornice and the machine was used by the Italians during the Abyssinian war. Two years later in 1939 England found herself at war again and William Tritton was commissioned to design a brace of experimental tanks. Lincoln had again become 'tank town' with the larger manufacturers such as Ruston and Hornsbys and Ruston Bucyrus

A Second World War and a second generation of Germans get acquainted with a Foster designed tank twenty years after it was supposed to be obsolete

Foster's Tog I being evaluated on the Test Ground in late 1941

The imposing TOG 2 on Fosters testing ground in Lincoln early in 1942

soon going into full swing producing Matilda, Crusader and Cavalier tanks. However, many of the army's top brass had seen at first hand the horror of the trenches 20 years earlier and they were worried that the next war could end up being fought in the same way. The huge tanks designed by Fosters were built with this problem in mind and were therefore quite unsuitable for the urban warfare or progressive countryside battles that would eventually come to characterise the tank engagements of the Second World War. The two tanks were christened TOG 1 and TOG 2. The name TOG stood for 'the old gang' reflecting the distinguished team of tank pioneers responsible for the new tanks creation, Tritton, Wilson, Rigby, Stern and Ricardo being just some of the names involved. The two machines were very different from each other, each being created to test individual drive system designs. The engine of TOG 1 drove an electric generator that transmitted power to each of the tracks. TOG 2 drove the tracks through a hydraulic system similar to that used in the Mk VII of 1917. Neither of the TOGs saw action during World War Two and only the second of the two tanks still survives. At 70 tons (71,125kg) TOG 2 has the dubious honour of being the heaviest tank in the Bovington collection, so heavy in fact was the tank that few bridges in Europe would have been capable of holding it had it ever seen action.

The massive TOG 2 preserved at the Tank Museum, Bovington

Virtually nothing remains today of the Lincoln factory where the massive TOGS and the worlds first tanks emerged. William Foster and Co LTD continued to produce engines, threshing machines and pumps until they finally went out of business in the 1960s and ironically became swallowed up by the surviving company of Gwynns Pumps. The last steam traction engine produced at Fosters works was dispatched in 1942 and is thought to be the last steam powered traction engine to be produced in a British factory. This is perhaps not an achievement to be proud of. If they had embraced change and expanded into petrol, diesel or hydraulically operated machine production they may have survived for longer. Despite this, Tritton was so happy with the companies achievements during the war that the tank became the official trademark of William Foster and Co LTD. Every product produced after 1918 and even the company letterheads carried the tank logo, in fact Fosters were so proud of the tanks that they even created the famous Tank Plaques. These were just over 3 ft (92cm) in length, made from cast iron and rather oddly depicted a Mk I tank without its steering gear. The plaques were sent to companies and agencies who had assisted in the creation of the first machines. An example sold recently in a Lincoln auction room for over £600, although as with so many other things fakes are now starting to appear on the market.

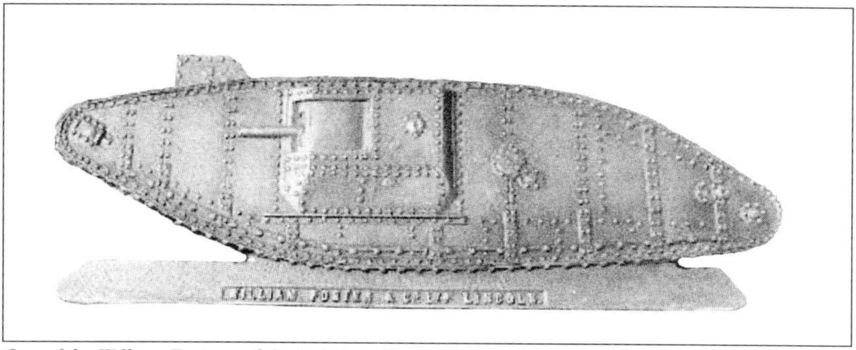

One of the William Foster and Co Ltd 'Tank Plaques'

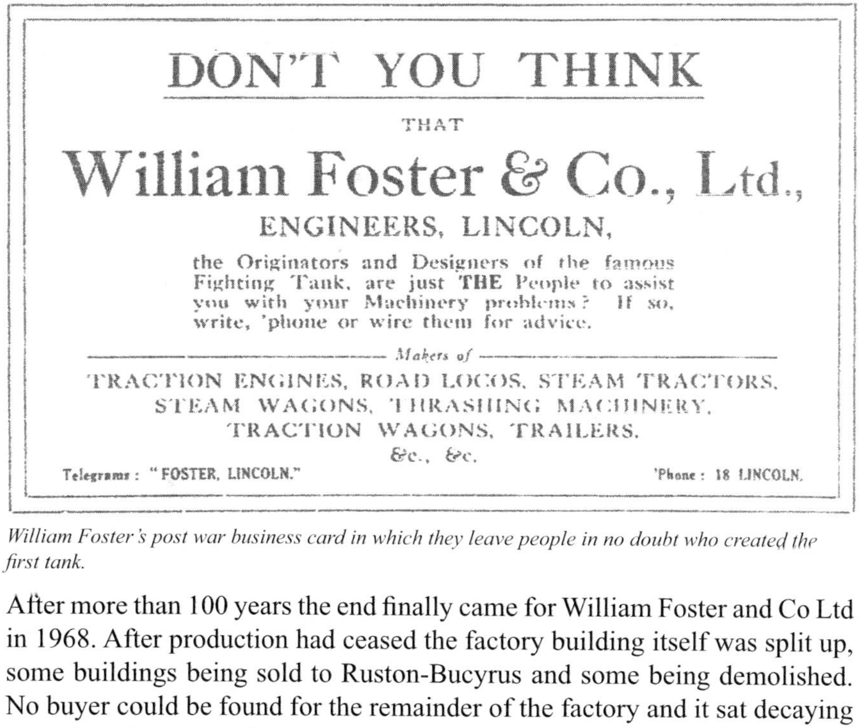

William Foster's post war business card in which they leave people in no doubt who created the first tank.

After more than 100 years the end finally came for William Foster and Co Ltd in 1968. After production had ceased the factory building itself was split up, some buildings being sold to Ruston-Bucyrus and some being demolished. No buyer could be found for the remainder of the factory and it sat decaying until the mid 1980s when an out of town shopping centre was planned. The last factory buildings were flattened and Tritton Road was created to serve the new shops and remind future generations of Lincolns once proud industrial past. The Lincoln Tank Group managed to save several irreplaceable Foster artifacts including the huge carved stone factory name plaque, which is now housed next to Flirt II in the Museum of Lincolnshire Life on Burton Road, Lincoln. The last remaining in-situ fragment of the factory can still be seen abutting the north side of a row of early 20th century terraced houses. This plain, brick built, buttressed wall is not impressive and thousands of people drive past it everyday without giving it a second glance, but this structural fragment is the south wall of Fosters machine erecting shops. It was in these erecting shops that 101 Mk IV tanks and 200 Whippets were created during 1917 and 1918.

Although the factory may be long gone, at least the company name lives on Dawson Bros of Bicker, Lincolnshire have bought the rights to the William Foster and Co Ltd name and just finished their first brand new Foster single cylinder Wellington Tractor.

THE GREAT UNKNOWN
JOHN (to unknown inventor): "Tanks."

Another Lincoln building that has survived slightly better than the factory and was also pivotal to the creation of the tanks is Bracebridge Hall situated at the crossroads of Newark Road and Brant Road. In 1914 William Foster and Co LTD employed a full time staff of 350 people, but by 1918 more than 2000 men and women were working at their Firth Road premises. New employees were arriving almost every day from all over the country and at first they were expected to find their own accommodation. However, this soon became difficult as several other local factories were also trying to house an ever-expanding work force. In November 1917 a great mass meeting was held in Lincolns Corn Exchange in order to try to find a solution to the problem. The meeting was chaired by Col. Joseph Ruston who was supported by William Tritton. The meeting started at 8pm and went on late into the night without a long-term solution being found. William Tritton appealed to his established workers to rent out any spare rooms they might have in their own homes but by this time it was becoming obvious that housing Lincolns temporary work force was a logistics problem that required a drastic solution. Therefore, early in 1918 Fosters bought many houses in Waterloo Road, Coulson Road and Bracebridge Hall on the outskirts of Lincoln. The hall and its four acres of landscaped grounds suited the companies problems perfectly and on the 2nd March 1918 General Sir William Robertson officially opened the grandly titled 'William Foster and Co LTD Residential Club for Homeless Munition Workers'. Sir William Robert Robertson, 1st Baronet of Beaconsfield

Bracebridge Hall as it is today

GCB GCMG GCVO DSO LLD DL, known as 'Wully' to his friends, was a fascinating man with an equally fascinating military career. He was a Lincolnshire lad having been born at Welbourn about 20 miles south of Lincoln on January 29th 1860 although for some reason his official Army records show his birthday as 14th September 1859. On the 13th November 1877 at the age of 17, he enlisted as a private in the Worcester regiment and by March 1920 at just 60 years old he was a Field Marshal. He remains the only man ever to start as the lowest rank in the British Army and end his career as the highest. The opening of Bracebridge Hall in 1918 may not have been much of a high point in Robertsons career, but to those who would eventually live there it was undoubtedly a palace compared to the homes of most who came to work at Fosters during the war years. The hall included spacious dormitories, two billiard rooms, a music room, a games room, a reading room, a refreshment bar and several bathrooms, all of which had running hot water. When not building tanks or playing billiards the residents could walk the grounds, play bowls or grow their own fresh produce in the well equipped vegetable gardens. For workers who often worked more than 12 hours a day, Bracebridge Hall would have provided vital rest and relaxation, without which the quality and output of the War Materials they were producing would surely have suffered. After the war the hall became a Social Club for Lincoln workers in general, while still being owned by Fosters and was mainly intended for their employees. Eventually the need for temporary workers housing ended and Lincolns engineering prowess faded to be replaced by tourism and the new University. Bracebridge Hall ultimately closed as a workers social club and was eventually taken over as offices for North Kesteven Council. The hall has recently had another change of owner and is now the Grosvener Hall registered Nursing Home. Bracebridge Hall has survived into the 21st century, purely because it has continued to fulfil a useful role in Lincolns society and is now a resplendent Grade 1 listed building. Whereas Bracebridge Hall has continued to change and accept re-use, the majority of Lincolns other landmarks in the story of the tanks development have not faired so well. Much the same fate that befell Fosters main factory building has also left Fosters tank testing ground mostly covered with modern out of town superstores. Much of Britain's rich industrial and social history is now represented by proxy with road names, motorway service stations and blue plaques. Fosters is no exception, for the factory and its glorious past there is Tritton Road and for the testing ground there is a blue plaque in the foyer of Morrisons supermarket that reads,

Sir William Tritton 1875-1946.
Near this site in 1915 Sir William Tritton,
Managing Director of Wm. Foster and Co, agricultural engineers,
designed, built and tested the first prototype fighting tank.
The prototype, 'Little Willie' and its successor 'Mother' ended the
stalemate of trench warfare and led to allied victory in World War 1.

Foster products including Whippets, tractors and a Mark IV crowd the testing ground in late 1918 (The Tank Museum, Bovington)

While Tritton was ensuring the companies place in the history of armoured fighting vehicles by creating monsters like the TOGs, in France they were trying to keep the might of the German army at bay for a second time in less than thirty years, by using the same light tanks they had first designed in 1917. Although the backbone of the French Tank Corps was the very capable and well designed Char-B, the First World War Renault FT-17 was such a simple and useful machine that even in 1940 the French were still fielding the same basically unchanged tank. However good the little Renault may have been against machine guns and infantry in 1917, by 1940 the crews were committing virtual suicide by engaging an enemy equipped with the improved Panzer IV and 88mm field artillery that could reduce a tank to nuts and bolts with a single shot. Despite the odds against their survival several Renault FT-17 light tanks are still known to exist around the world as they proved very popular with not only the French, but also with the Americans and the Japanese. The FT-17 was in service with several countries armed forces right up until the outbreak of the Second World War. Examples can still be seen at the Bovington Tank Museum, the German Technological Museum in Cologne and the Musee de Royale de l'armee in Belgium, to name just three. Perhaps the rarest of the first generation of tanks are the

A male Renault FT17 exhibited at the Musee de Royale de l'armee, Brussels

German machines. Although the Germans did create other tank designs, such as the Leichterkampfwagon, their version of the Whippet, and the fearsome, but totally impracticable, 100 ton (108,865kg) monster known as the K-Wagon, only the A7V Sturmpanzerkampfwagon would see action during the war. A total of just 20 A7Vs were finished before the armistice. Many countries including England, America and France had examples of captured A7Vs at the end of hostilities in 1919. Unfortunately, as with their British counterparts, the importance of the machines was not recognised and one after another they were scrapped. Two A7Vs, Hagen and Schnuck, which had been captured by the British in 1918, were displayed on Horse Guards Parade until the early 1920s, when they too were also scrapped. Only the 57mm cannon from Schnuck still survives and is kept at the Imperial War Museum, London. The American owned example, Nixe II, did comparatively well and survived until 1942 when it was cut up in a symbolic gesture against Nazi Germany.

The sole surviving A7V, 506 Mephisto, was captured by the Australians at Fremicourt in 1918 and shipped back to Brisbane as a war trophy. The tank was nearly lost in 1946 when the Museums acting director called for its removal and simply dismissed it as '*an ugly object*'. Undoubtedly the A7V is an ugly object and was never going to win any design awards, but it is unique, perhaps the rarest tank in the world and thankfully it was not removed.

One of the few, A7V Hagen, on Horse Guards Parade shortly before being cut up for scrap

Thankfully, attitudes towards the Australian owned tank have changed since 1946 and A7V Mephisto has recently been fully restored. The tank is now on display at the Queensland Museum in Brisbane. The German Panzer Museum in Munster recently approached the Queensland Museum in an attempt to negotiate the return of their tank, but when their offer was refused they decided instead to build a fully authentic working replica from original drawings. Their replica is a remarkable interpretation and is indistinguishable from the real thing. It now stands in pride of place in the museum next to their example of the equally rare and original Leichterkampfwagon.

The rarity of the first tanks has not stopped some avid and wealthy collectors from adding one to their collection. The New Zealand based film director Peter Jackson responsible for cinema epics such as 'Master and Commander' and 'The Lord of the Rings' trilogy is the proud owner of a full scale working replica Mk IV female tank which he regularly displays at shows in his homeland. A wealthy Leicestershire businessman and military historian is currently trying to locate as many original drawings as possible with an aim to producing an exact full scale working replica here in England. The Belgian town of Poelcapelle had a Male Mk IV on display in the main square until it was removed by the occupying German army during the Second World

Now the Tank Group of the Poelcapelle 1917 Association propose to reinstate the towns tank, either by excavating one from the battlefield or by making a replica. In 2003 a new group called 'The Friends of the Lincoln Tank' was started with the aim of publicising Lincolns accomplishments during the Great War. In addition to research and organizing lectures and tank related events the group intends to raise sufficient funds to create a working replica Medium A Whippet or a Mark IV tank and several members of the group are currently practising by working on scratch built radio controlled models of both Mk IV and Whippet machines ranging from 1/16th right up to 1/4 scale.

Whatever their nationality, survival rate or state of preservation the few original Great War tanks that still survive have one thing in common. They are all lasting and tangible monuments to the brave men who fought for their country using pioneering and untested technology knowing all too well that there was a very good chance that they might never return home. In addition to honouring their memory, the surviving machines are a permanent reminder of the people who had to strive, not to create a more efficient killing machine, but to end the horror and find an alternative, their ideas and sketched designs culminated in the tank and Lincoln should be proud of it's achievements.

APPENDIX 1

Taped reminiscences of William Foster and Co Ltd employees Undertaken by the 'Lincoln Tank Group' between 1977 and 1984

The workers at William Foster and Co Ltd did valuable work during the First World War, but just as valuable as their engineering contribution was the fact that they knew when to keep quiet. Their secret work in Lincoln stayed a secret and they, and their allegiances were never in question. Many other factories employed on War Work had trouble with their employees that ranged from idle gossip to downright espionage. For instance the Metropolitan Wagon and Finance Company in Birmingham were also engaged in building tanks and several of their workers were arrested for '*loose talk'* in the pub. Twelve workers at a Swiss based company associated with the Birmingham factory were questioned by the Special Branch for breech of the Official Secrets Act, one man being sent to an internment camp for the remainder of the war. With the armistice in 1918 the secrecy ended and Fosters full part in the tank story was made public, Tritton and Wilson became household names. Early in 1920 the official William Foster book *'The Tank; It's birth and Development'* was published. The book told the whole tank story from the firms point of view. It was well written and lavishly illustrated with special leather bound and gold embossed editions going to companies and individuals of special note. Exclusive leather bound copies were produced specially for the star players such as Tritton, Wilson and Stern that had their initials embossed on the front cover. For all its splendour the book was obviously quite biased towards the Foster, Tritton and Wilson side of the tale, with little mention made of the workers and in a way overshadowed the individual stories. Later generations forgot what had happened during the Great War or were simply not interested in Grandmas rambling tales. As is the case with primary oral history, the people in question get old and eventually die and the everyday anecdotes of hardship and camaraderie are lost forever. The grand official histories of the Great War told thrilling tales of the first day of the Somme or the deeds of the Red Baron, but with every day that passed the ordinary people who had been there at the time and done their bit, at home or in the trenches, were getting fewer and fewer. The incredibly important 'run of the mill' details would soon be gone for good. With this in mind in the mid 1980s a band of history minded enthusiasts got together to form the Lincoln Tank Group. Together they compiled and published the renowned 'Lincoln Tank Papers', now sadly long since out of print, and helped to secure the female Mk IV tank Flirt II for permanent display at the Museum of Lincolnshire Life. With great foresight they also spent an equal amount of their time and resources travelling all over the country recording the personal reminiscences of any surviving William Foster and Co Ltd employees or First World War Tank Corps veterans that they could locate. Even in the 1980s this was not much more than a handful of people and even they have now all passed away, leaving these few precious interviews as an irreplaceable piece of social history, a window into their working lives of over 90 years ago. For the first time the workers got to tell their own tales of everyday life and the following chapter is in their own words. It has been taken from the old reel to reel tape recordings kindly supplied by the collections custodian and renowned Lincolnshire historian Mr. Ray Hooley.

Two William Foster and Co Ltd Munitionettes driving an engine and wagon in the works yard

Mr. William Rigby
Chief Draftsman and Designer at William Foster and Co Ltd
Interviewed in late 1977 when aged 89

The whole thing started with Mr. William Trittons visit to South America with an oil engined traction engine in 1913. We also made another tractor with tracks, Hornsbys built the track and it was sent to Alaska, but I have heard that water got in between the tracks and turned to ice and the whole thing collapsed? We then built a large Daimler petrol engined tractor that was used for hauling parts of guns for the defence of Paris. After that the machine, this tractor was used to cross a trench by means of a bridge that it was carrying. The tank developed by means of a track, originally bought from America, it was no good and had to be redesigned and rebuilt in Lincoln. The reason it was called a tank was that the boilermakers got so fed up with alterations that they called it the bloody tank instead of a water carrier for Mesopotamia. It all developed from this tracked engine we made for Brazil and then Sir William Tritton and Major Wilson got together and from their combined knowledge we eventually made the tank. There were no leaks from factory and even the people at the works only found out what the thing was after it had been used in war. Wilson designed the outside shape, but all of its insides, the engine and all that, came from this Traction engine we'd made to haul guns for the defence of Paris. I remember the day that the first tank left Fosters factory, in fact I took it out. I was chief designer, well, draftsman I suppose you call me, Charlie Maughan was head test driver, but I had to show him how to work the thing. During the war there were about 2500 tanks made and of that Fosters probably made around 500, various other people were soon brought into it, but the tanks were always made from Fosters designs. We bought Bracebridge Hall to use as a sort of hostel for our erectors and fitters who came here to work for us. I used to work six in the morning until five at night, but if we were busy we would work all Friday night and not stop until twelve on Saturday afternoon, they were good old days. I was on 26 shillings a week when I had qualified. We got just one week holiday a year for trips week, perhaps a couple of days at Christmas and maybe one day at Easter. When Fosters had finish with the tanks it was a bit of a blow

Just a few of the ordinary faces behind an extraordinary invention. The staff of the riveting shop taken in late 1917.

as we had nothing to do, we'd lost all our trade to America, so we started on showmans steam traction engines and threshing machines, but we didn't do very well. Gwynns had gone bankrupt after trying to get into the motorcar trade and so Sir. William, who had been apprentice at Gwynns bought the company for Fosters. After Fosters had bought Gwynns the shares went up by 18 times. We really started on pumps for power stations, docks and gravel installations. I enjoyed the land drainage and pump side of things and went all over the world. I went to New Zealand to see some land drain stuff and power stations, we also sold a few in South Africa. In 1937 Fosters designed and built a tank for Italy and I went out to the Grand Cornice to test it. It was not a development of the old tanks, it was something quite new, two Italians came over to the works and the whole thing was put under my control. It was used in the Abyssinian war. Me and my daughter went out to Venice just before this and I took an order for a 2' 6" threshing machine for Italy, they are usually 4' 6". Then the Abyssinian war started and we were told that if we didn't get out soon we'd not be able too, so we left quick. In the last war we built two huge tanks that are now in Bovington. They also have Big and Little Willie and several others, so we are well represented there.

Mrs. Dorothy May Hare
Machinist at William Foster and Co Ltd
Interview March 1984 when aged 84

In 1915 I was 15 years old and I started work at Fosters. I started on a simple, single drilling machine. I ended up on a double driller and I learnt how to use all of them, if anybody was off I used to take their place. I used to make the shoes that were underneath the chain on the tank track. I drilled holes in them and shaped them on a plainer. They were rough cast steel, you could cut yourself on them. They were sent to another department for finishing after that. That's where I was when the siren went, I'd stacked all the shoes up and clean forgot, I switched my machine off and all the lights went out. I turned around and went straight into the pile of shoes. They had to come and dig me out, I got cut a bit, but it wasn't hospital treatment. I once got my overall sleeve caught in the driller, it tore everything off me! Mr. Walker, the charge hand, took the belts off the pulleys, and that stopped all the machines, I wasn't hurt. You could make as many as you liked, it wasn't peace-work, some was ever so idle, but I just carried on. It was all girls, no men except the charge hands. They never left the room, they always had their eyes on us. We all wore blue overall and bonnets while we were working, some had the Foster works name on them. We would finish at six at night and the night people would take over, then we would take over from them again at six in the morning, the machines were never switched off. Sometimes I was there until nine or ten at night. We would get extra pay for Sundays and night work, I started on 15 shillings a week and built up to about 25 shillings by the end.

Women workers busy at their drills with a stack of brand new track shoes in the centre

We lived on Coulson Road just next to Fosters, so I could come in through the back gate. The houses on Coulson Road belonged to Fosters factory. I went to the tank testing ground one day, we shouldn't have been there, but there was a gentleman I knew, Mr. Maughan. He said *'would you like a ride'* and I said yes, I was only in about five minutes and I was standing on a little platform at the side. The testing ground was on the old Pennistons Field. I only met Mr. Tritton once, out on Lincoln Racecourse, he was with Lady Tritton. The King and Queen came to visit the works once, but I missed it as I was on nights that week.

I left Fosters in 1918 when the men started to come back to their prewar jobs, I was the first to give up my job. I had a good time at Fosters, it was hard work, but I put my heart into it, all us girls got on very well together.

Mrs. Preistley
Machinist at William Foster and Co Ltd
Interviewed March 1984 when aged 85

I had always wanted to be a dressmaker, but you had to wait so long before you could get your apprenticeship, the war broke out and put a stop to all

Rough cast track shoes being shaped on the planer

that, so I went to work at Fosters when I was 15. I worked on a lathe in the machine shop making shoes for the tank tracks. We were supplied with hats and overalls, you had to wear the hats to stop your hair getting caught in the overhead belts. I lived on Waterloo Street, where Fosters back fence was, they owned all the houses on the street. We had a good look around a tank one day, but I never had a ride in one, I just did the track shoes all the time I was working there. We didn't get many holidays, it was work, work, work, not like it is now. The finished tanks used to go up Waterloo Street, over the tank bridge that was specially built and then onto the testing ground.

I met Sir William Tritton several times, he seemed to be a nice chap, pleasant enough. The Fosters men who didn't go off to war had a little enamel badge so that people would know they weren't cowards, Rustons did the same. All of us girls were finished in 1918 when the men came back.

Mr. George Atkin
Dispatcher at William Foster and Co Ltd
Interviewed September 1984 when aged 88

I began working at William Fosters in 1911. We used to have to write labels and make out consignment notes for the carrier companies, everything went by rail in those days. We even sent the traction engines and threshing machines by rail. I remember that one day there was a trench cut in the main yard and this Foster petrol engined tractor was made so that it could cross trench and pick its bridge up again afterwards. Major Wilson came down and he worked in our drawing office. Then we got these caterpillar tracks sent to us and they were attached to the machine we called Little Willie. When the machine called Big Willie was built it was made differently so that the tracks went all around the machine I think this was Major Wilsons idea, but I'm not too sure. Once these machines had been used the Germans started to make their trenches wider, so we made tanks that we called Tadpole Tails, this made the machine and the track longer. Soon the Germans started to use nickel headed bullets that went straight though the 3/8th (inch) thick plate on the side of the tank. A man came from Osbourns of Sheffield selling armour plate and Mr. Tritton came down to see him. The plate was only about 1/8th thick. The idea was to make breastplates for chaps to wear under their tunics. Sir William said to him *'Lets try this plate of yours out then, would you like to stand behind it?'* He didn't, but this chaps armour plate was put up against the wall with one of our standard 3/8th thick pieces in front of that. We fired a bullet at it and the bullet went straight through the 3/8th and the new 1/8th piece. Sir. William said *'It's a good job you didn't have too much confidence in your product'*. We didn't tell this chap that we had used a German nickel nosed bullet and he went away without ever knowing. He was a good man, Mr. Tritton, but you knew he was the boss, if you know what I mean. I never really spoke to Mr. Wilson, he seemed a quiet and very busy sort of a man.

Mr. Quincy, Mr. Rose and Mr. Maughan were the testers who drove the first tanks, I think it was Mr. Quincy who drove a tank with the King in it once. He was having trouble getting to one of the levers and had to ask the King to *'utch up a bit'*. Mr. Quincy once set up a stunt where he was run over by a tank on the Fosters testing ground. He actually fell to the ground between the track frames and the tank went over him, from the side it looked like he'd been flattened by it. You'd need a good driver to help you with a stunt like that, he probably used Mr. Maughan to help him. He was a bit of a clown Mr. Quincy, his sons were the heads of some colleges in Australia or somewhere, they must have got their brains from Mrs. Quincy.

I became a buyer and used to get small firms involved in making things like track links and pins. We were never short of anything we needed for making the first tanks. In 1918 the armistice came as a complete surprise, the military work at Fosters stopped and we went back to steam traction engines and tractors just as we had been making before the war. We did very nicely from the money we'd made due to military contracts, but later on we shipped twenty sets of threshing machine and traction engines out to Rumania. The company never got paid and that's where Fosters lost a great deal of their money, these were something like about £500 per set, maybe more. As a result the company had to issue debenture shares, which made things very difficult. Once companies found we'd issued debentures, they were soon after their money, and that was money we didn't have, which made things worse.

During the war we'd been given armbands and a letter to show that we were employed at Fosters on War Work. This was to stop people on the street giving you white feathers for cowardice. I remember this one lad who worked in the offices at Fosters, Keyworth was his name, he joined up and got the Victoria Cross during the war.

Lance Corporal Leonard Keyworth V.C.

Leonard Keyworth was born in Lincoln in 1893 and attended Rosemary Lane School in the city. In August 1914, at the age of 21 Keyworth tried to join the Lincolnshire Regiment but was not accepted because he was in a reserved occupation at William Fosters. Still keen to enlist, in September 1914, he left Fosters and went to London and joined the 1/24th (County of London) Battalion, the London Regiment. Keyworth won his V.C. for an action at Givenchy in May 1915. On the 3rd July 1915 the London Gazette wrote that he had won the medal for 'most conspicuous bravery at Givenchy on the night of 25th - 26th May 1915. After the successful assault on the German position by the 24th Battalion, London Regiment, efforts were made by that Unit to follow up their success by a bomb-attack, during the progress of which 58 men out of a total of 75 became casualties. During this very fierce encounter Lance Corporal Keyworth stood fully exposed for two hours on the top of the enemy's parapet, and threw about 150 bombs amongst the Germans, who were only a few yards away'. Leonard Keyworth was recommended for the Distinguished Conduct Medal by his company commander, but was actually awarded the Victoria Cross. The first he knew of his VC award was when he read about it in a newspaper containing the citation on 4th July 1915, he received it at Buckingham Palace on the 12th July 1915, but was killed by a sniper's bullet just three months later on the 19th October 1915.

Mrs. G. Atkin
Typist in the Dispatch Office at William Foster and Co Ltd
Interviewed September 1984 when aged 85
I started by working at Fosters in 1917, I was checking invoices in the boardroom. They found that I could do short-hand and so I ended up as a typist. I used to send out the money to the relatives from the William Foster Comforts Fund. We worked from six in the morning until half five at night, my father was a foreman at Ruston and Proctors and he used to say that Fosters was like a convalescent home.

I never saw much of Mr. Tritton and I can't remember meeting Major Wilson either, but I do remember standing on an overhead gallery and watching the King and Queen walk through the erecting shed when they visited the works in 1917. When the armistice was declared we didn't have any special works celebrations, but I think we got the day off and there was a service at St. Peter and Gowts Church. I enjoyed working at Fosters and was very sorry to have to leave, but the male staff members were coming back from the army, the men were being demobbed so I had to go, Joe Rawling took my job.

Mrs. Westwood (Nee Broadberry)
Grand-daughter of Thomas Broadberry Boilermaker at William Foster and Co Ltd - Interviewed September 1984 when aged 70
My Grandfather worked at Fosters for all his life and was put on this new project in 1916 that turned out to be the first tanks. He would have been 55 in 1916, I can just remember him he died in 1924 when I was ten years old. He had a solid silver badge shaped exactly like a tank, a First World War tank, with all the armour and the studs on it. It was about two inches long and was always kept in a little box in my mothers drawer. On the back of the badge was engraved *'Presented to Thomas Broadberry by William Tritton'*, I can't remember the date, but my mother was never one to treasure old things, so I've no idea what became of it. After the first war, I used to go into Fosters quite often, my grandmother would send me with a little wicker basket of hot food for my grandfather. I don't remember seeing any tanks and I don't really know anything else much about it, I just remember my grandfather's silver tank badge.

J.R. 'Bob' Burley
Apprentice Fitter and Erecter at William Foster and Co Ltd
Interviewed May 1984 when aged 83
I was thrilled to start work at Fosters as I was very interested in the tanks and wanted to know more about them. We started our day at six in the morning,

then breakfast was eight till half past, and we often used to work extra hours, usually until eight in the evening. We often had a speaker come and talk to us during the breakfast break and we used to give them a reception and bang on the tables. I'd read all about the tanks before I went to Fosters, there were no pictures of the tanks in the papers and I'd never seen one until I went to work at Fosters. There was no production line, they would just erect one tank in a space and it would stay there until it was finished. All of the plates, the shell, was riveted together in another department and then it would come to us, where we would fit the engine, the gears, the sprockets, tracks and so on until it was complete. I chiefly did detail work on the Mk IV like the controls and bearings for the rollers underneath the tracks. I didn't get involved with greasing the rollers, I didn't do any of the dirty work. After this the Mk IV was extended and then finally the Whippet came along with it's two Tylor engines. The Whippets were more manoeuvrable, the Mk IV was clumsy and heavy. With the old tanks there had to be one man on each track to put it in gear, but the Whippet was controlled by the different track speeds. We had our own testers, but we also had the Tank Corps working with us. We had nothing to do with armament at Lincoln, the tanks were fitted up with their guns elsewhere.

Mr. Tritton would often come and see me, he was a good fella to get on with but he demanded absolute perfection. He was a quiet sort of man, but you could imagine that his mind was working away all the time. We all called him *Tanky Bill* or *Old Bill*, but not to his face. Old Charlie took the King around the test field in a tank and afterwards everyone drifted off except me and Charlie shouted '*Get in I'll take you for a ride*' and I got into the Kings seat and that was my first ride in a tank. Charlie Maughan was the chief tester and if he tested anything, it had a real thorough test, he put it through it's paces. The testing ground was covered in holes that I think would compare with anything you'd find in France. I also remember General Robertson coming round the factory, we didn't seem to make any special arrangements or effort for them.

Mr. Rigby was a real family man, I never knew him in the tank days, but I knew him well after the war and I kept in touch with him right up until his death. I remember Billy Bell, who would later be Mayor of Lincoln. He was a ganger in those days and I remember he wanted some nuts tightening under a tank radiator, but he couldn't get at them. I was only small so I managed to crawl underneath and do the job, he gave me a shilling, which was a fortune to me in those days. After Cambrai in 1917 we'd lost a good few tanks and also people realised that the tanks was a valuable weapon, so this meant that we had to work even longer hours, but we wanted to do our best and help the lads in France. The testing field was alongside the railway and a concrete ramp was built so that the tanks could drive up and straight onto the railway trucks. The women workers at Fosters were employed mainly as crane drivers, or in the stores or as drillers, they were a great help. Everything always seemed to go very smoothly to me, but I suppose I was only concerned with the job I was doing. It was a great place to work, we always played jokes like sending the new apprentices to the blacksmiths shop for a bucket of steam or to the stores for a long stand. We all cycled or walked in those days, occasionally we'd go to the pictures or play football, but I was never very keen on dancing. We seemed to stop making tanks as soon as the armistice was declared and we went on to road locos, like the old Fair Ground engines with the big dynamos on the front. I stayed at Fosters until 1921 when trade was so bad that we were only working three days per fortnight. I couldn't stand that so I left and did 6 years in the RAF as an aero-engine fitter. In the mean time Fosters went from Steam engines to pumps and things got better, so after I came out of the RAF I went back to Fosters and carried on where I'd left off. In the Second World War I was on the reserve list and was called back into the RAF, so I don't know much about the Togs that Fosters made then.

Working at the vice in the centre foreground is a young Bob Burley when he was on the Whippet production line in early 1918

Chris Enderby
Pattern Maker at William Foster and Co Ltd
Interviewed October 1984 when aged 84

I started work at Fosters when I was 14 in 1915. At 14 years old I got five bob and worked a 54 hour week. I remember they built this tractor that would cross trenches. It wasn't armoured or anything. Little Willie was next, it was a trial tank, they tried it in the yard, I remember it running about. After the trials they built the proper one. I remember making patterns for the wheels at the back of the first tanks, we made a lot of those until they got rid of them from the tanks. They thought they could steer with them, but they were better off without them than with them. They had to build a bridge at the end of Waterloo Street so that the tanks could get from the works to the testing ground. There were lots of tank testers, like Mr. Quincy and Mr. Rose, but Charlie Maughan was the yard foreman. We also had a good few Navy and Tank Corps personnel there. I worked with Mr. Rigby, he was a young draftsman who ended up as one of the company directors. We had half a day off in 1917 when the King came around for a visit. The queen was with him, she always was, and they went round Rustons factory as well I think. Later on I worked on the Whippets, they were all completely different castings. As the war went on my money went up, I was apprentice for seven years. I left the company in 1922, but came back in 1927. In the second war we worked on the TOG tanks. I worked on them all night one night, straight through. That was before the Germans invaded France, so the TOGS never went into action.

Mrs. Connie Maltson (Nee Cownley)
Tracer in the Drawing Office at William Foster and Co Ltd
Interviewed October 1984 when aged 60

My father, J.T. Cownley, started work as an apprentice at Fosters and was very interested in the tanks and worked on them during the First World War. He helped Mr. Maughan when they were testing them in Lincoln, he eventually became a fitter and marker out. After the war he moved away from Fosters when work became short and he was on a three day week. He worked as a foreman for a company in Sleaford, then later on in the 1930s he came back to Foster and worked erecting pumps. I was always interested in his adventures and he would tell me stories about his work when I was a child. I went to work there in 1940, following my fathers footsteps. Later on in the Second World War he was involved in tanks again when they built the TOG 1 and 2. He used to test drive them with Mr. Maughan, but they realised that the tanks were extra big and they never went into production, but they did use them for various maneuvers and once my father had to go to Farnborough for a tank show, they usually have air shows there, but this was a big tank show. They took TOG to Farnborough and it was shown doing it's testing in front of some top army men. My father was a civilian and civilians were not allowed in Farnborough at the time so they had to borrow an army uniform for him and dress him up so that he looked as though he was a soldier. From the drawing office window we used to watch the TOGs being tested on the field at the end of Waterloo Street. The field was full of holes and ditches and we would watch in secret from the window.

I knew Mr. Tritton well, he was a happy looking man who always gave you a smile. He was very plump and genial, but he was the big boss so I never had any conversation with him except for saying good morning and that kind of thing. I only knew Mr. Maughan around the works and I don't remember lady Tritton at all, but Mr. Rigby was the boss of the Drawing Office, so he was my boss when I was there as a tracer. I used to work mainly on pumps, but when the Second World War came I did do some of the tank drawings, but I'm not an engineer, all I did was trace the paper drawings onto linen, I didn't know the details. Mr. Rigby was always very bronzed and always looked like he'd just come back from holiday. He was a very keen golfer so he spent a lot of time outdoors. He had twinkling blue eyes that showed up against his skin, he was a good boss, a very good boss. He used to somtimes send me out for his pipe tobacco or to order a haircut for him. He was always very busy and his time was limited, so if he wanted his hair cutting he would have no time to wait. He recommended my father for a medal, the B.E.M. and in his letter he said that my father had always worked very well with his whole heart in his work. It was a very happy place to work, everyone was always very friendly and nice.

Fosters Munitionettes in mid 1917

APPENDIX II

Drawings of Great War military vehicles produced in Lincolnshire's factories

Many of the weird and wonderful machines produced by William Foster and Co Ltd during the years of the Great War were never photographed. Several keen photographers were known to have been on their staff and the company had bought their own large plate camera some years before. It was quite usual for most companies to photograph their product range for inclusion in brochures and press releases, but in the case of some of Fosters creations they either remained unfinished or were deemed to be too secret to record. Machines such as the Flying Elephant or Fosters version of Major Hetheringtons 'Big Wheel' known as the Tritton Trench Tractor never progressed beyond the wooden mock-up stage. Consequently we are now only left with tantalizing sketches and snapshots of unfinished wooden structures to hint at the size and final appearance of these wonderful creations. Before Little Willie took on his final form he was known as the Lincoln Number 1 machine and was equipped with a turret mounted gun and the ill-fated Bullock Creeping Grip track set. For reasons of secrecy the only photographs taken of old number 1 show it covered with a William Foster tarpaulin. This seems odd when the lack of security surrounding the testing of the machines of Lincolns South Common is considered.

In the original first edition of the 'Landships of Lincoln' the author produced engineering line drawings of William Foster made Mk I and Mk IV tanks. In this second edition the intention is to expand on these drawings to include machines of interest produced by Lincolnshire firms such as Ruston and Proctor Ltd, Marshall Sons and Co and all of Fosters wartime military contracts including the unphotographed machines. As the book is intended to be enjoyed by enthusiasts of all ages and in order not to complicate matters the scale given for these drawings is an outline of William Tritton who in his safety boots and bowler hat stood at around six feet tall.

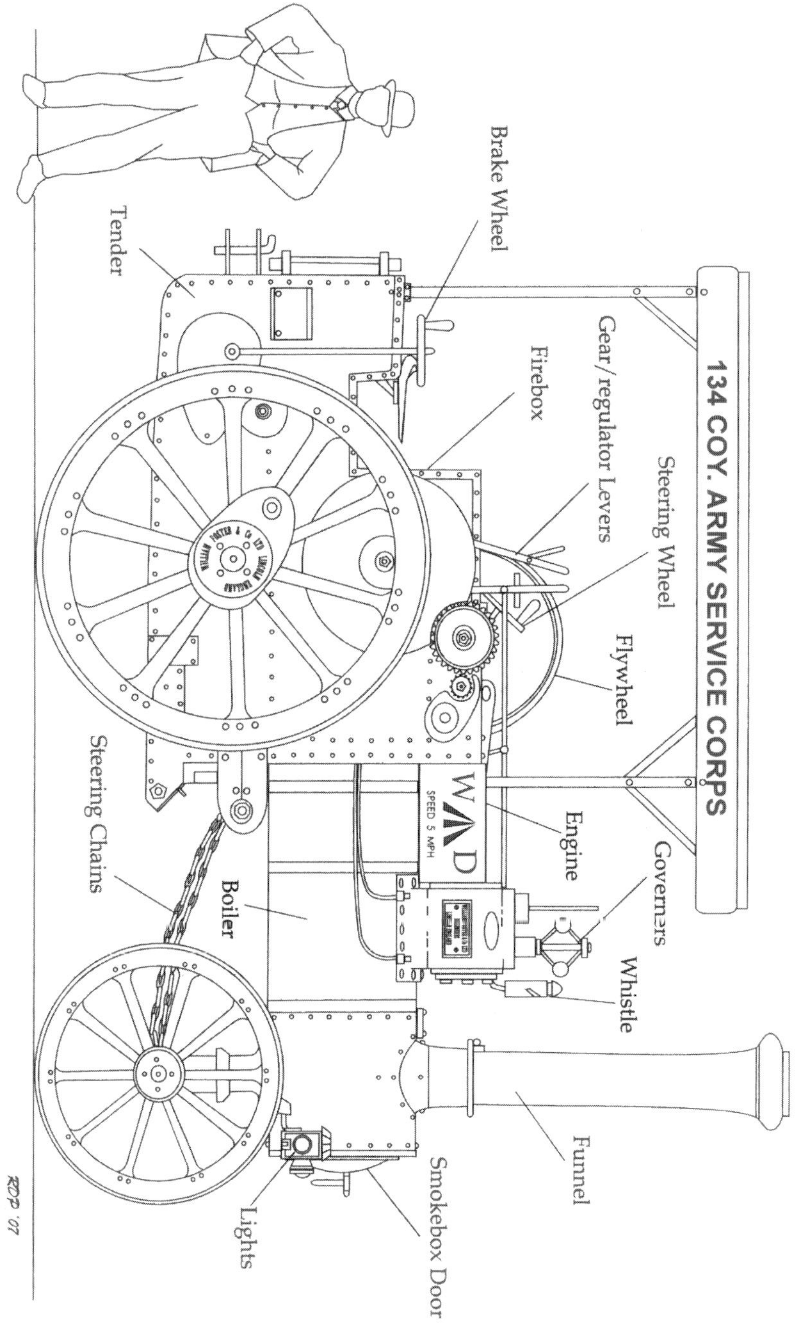

1907 William Foster and Co Ltd 'Wellington' Steam Tractor

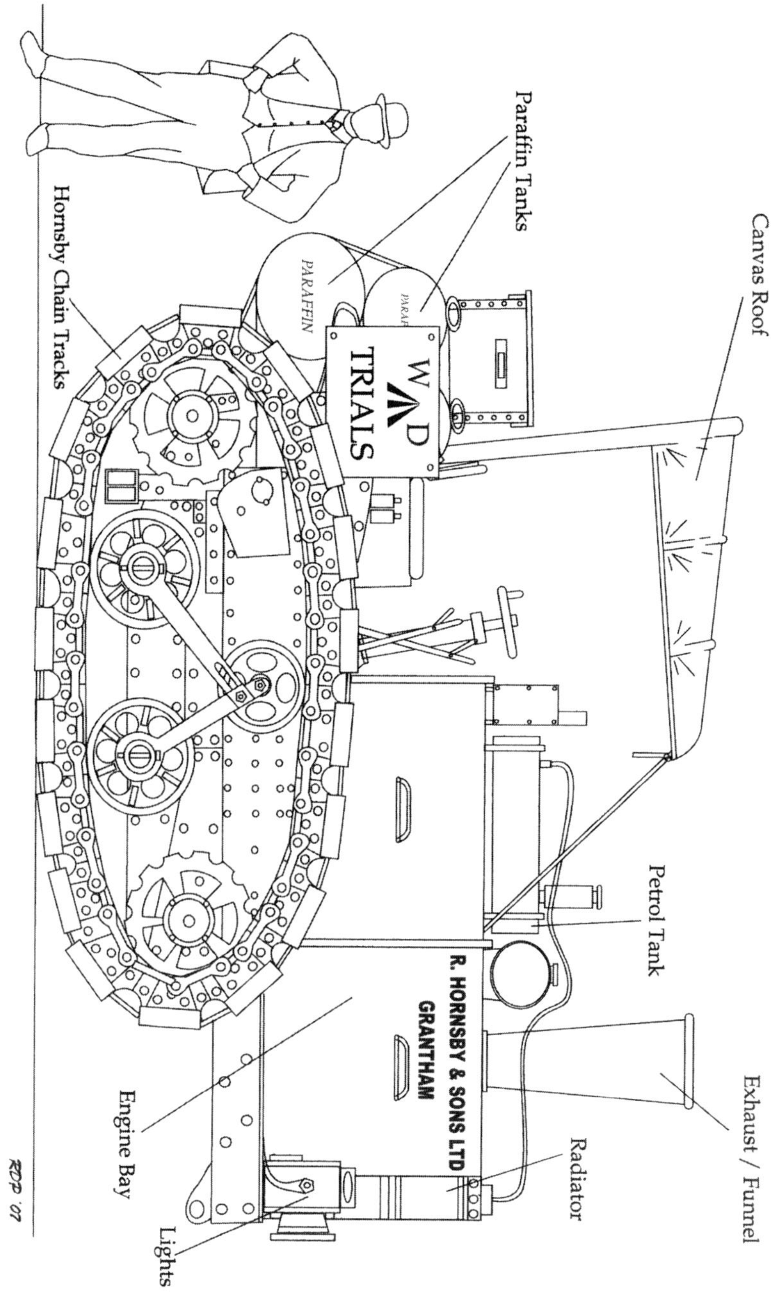

1909 Hornsby Petrol Chain Track Tractor

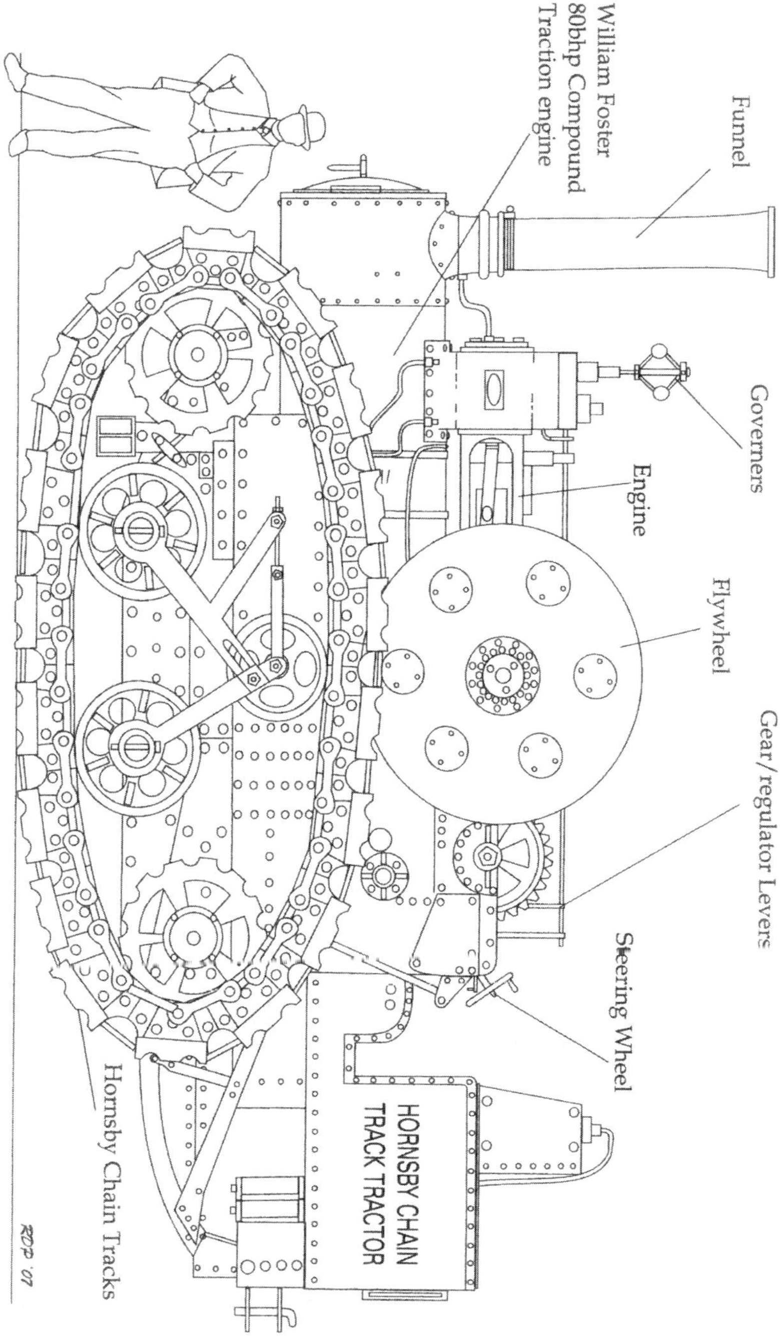

1910 Foster-Hornsby Chain Track Tractor

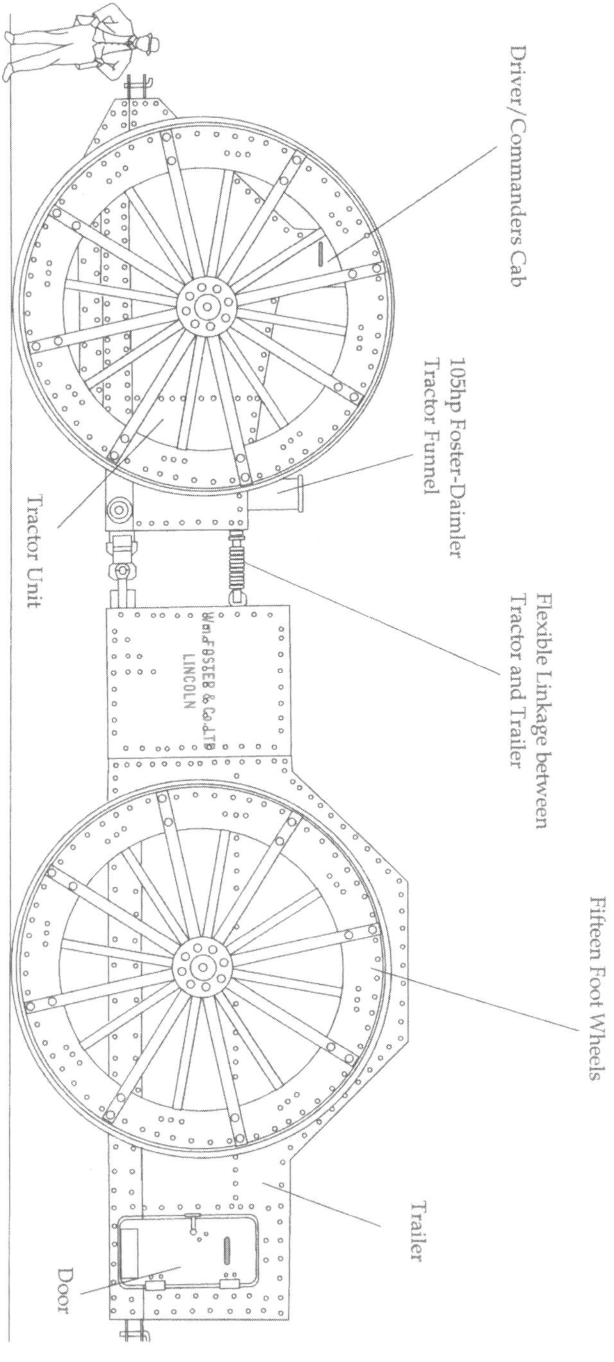

1915 William Foster and Co Ltd 'Tritton Trench Tractor'

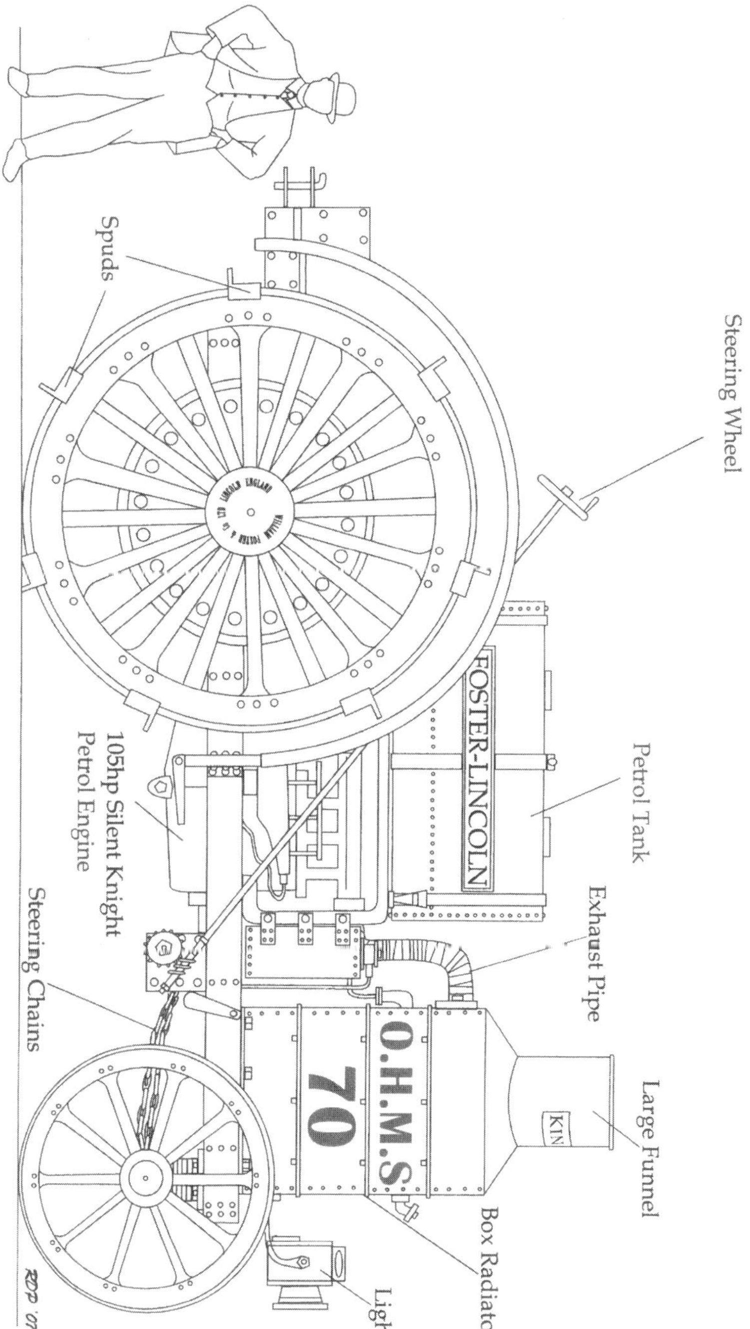

1915 William Foster and Co Ltd 105hp Foster-Daimler Tractor

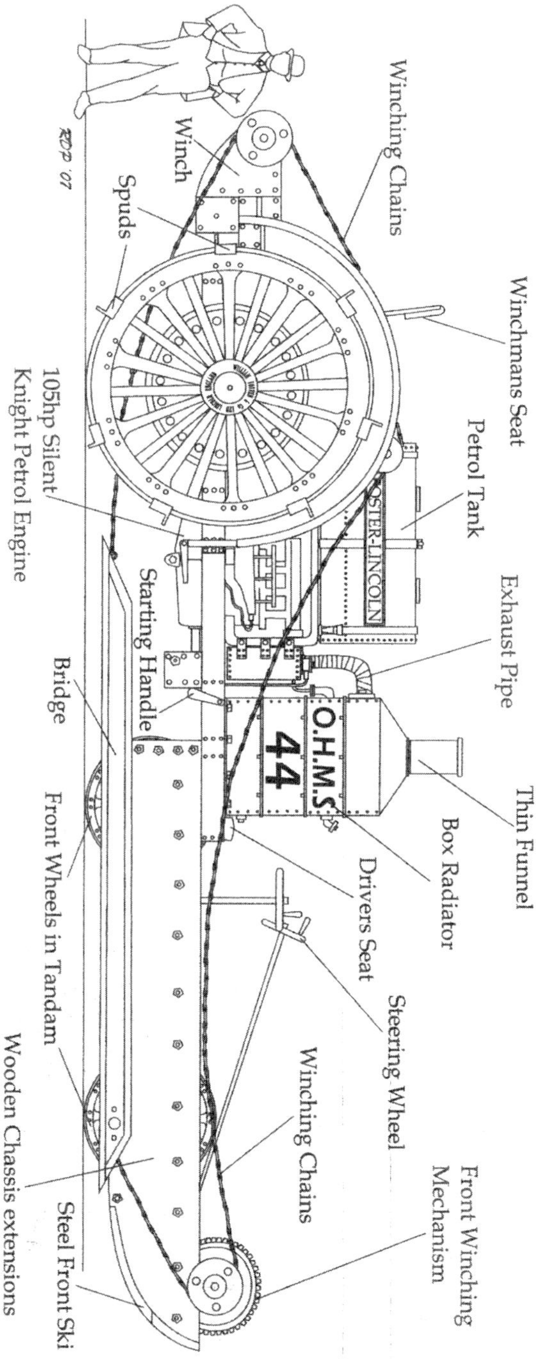

1915 William Foster and Co Ltd 105hp Trench Crossing Tractor

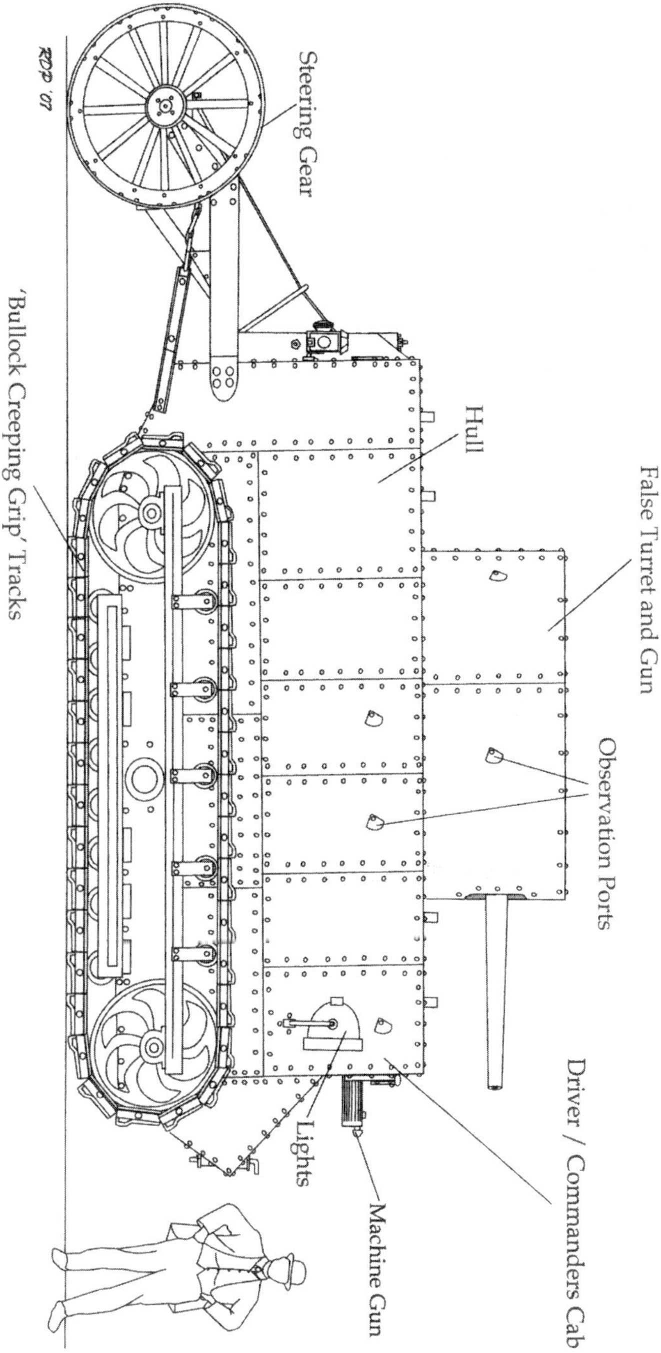

1915 William Foster and Co Ltd 'Lincoln Number 1 Machine'

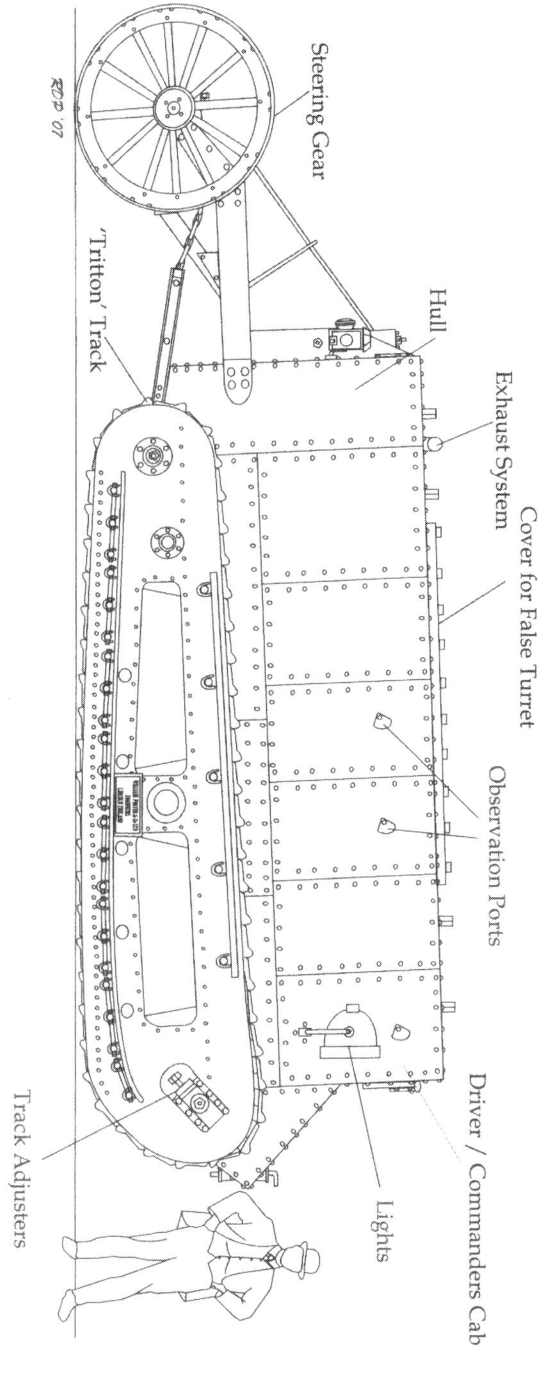

1915 William Foster and Co Ltd 'Little Willie'

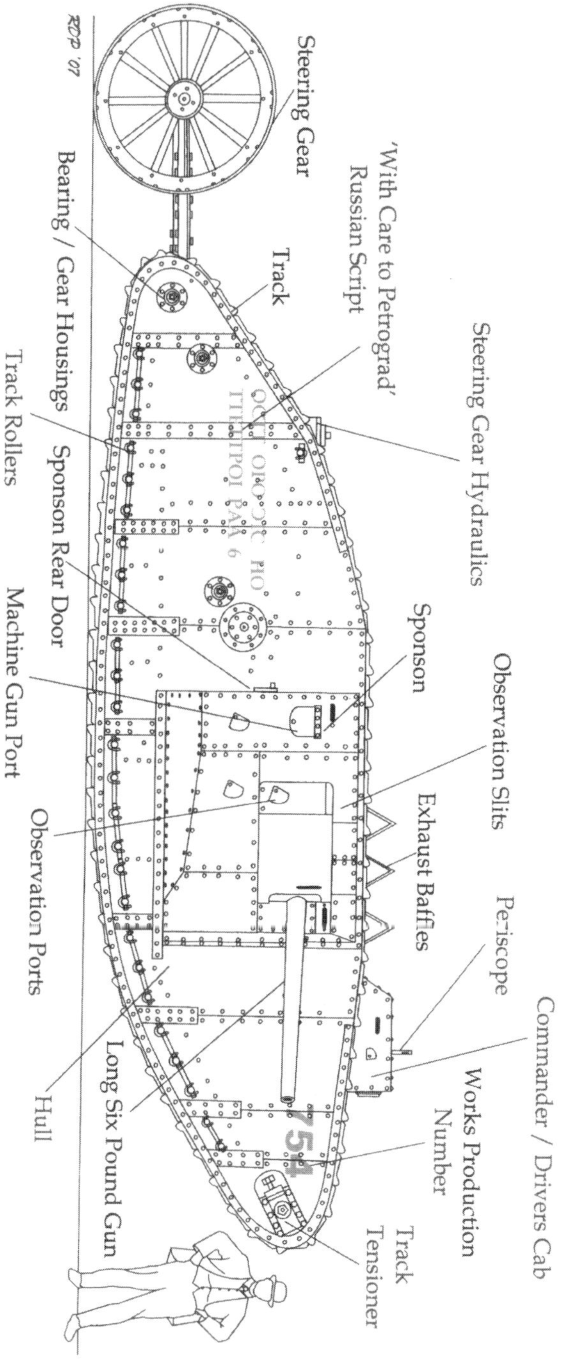

1916 William Foster and Co Ltd Male Mark I Tank

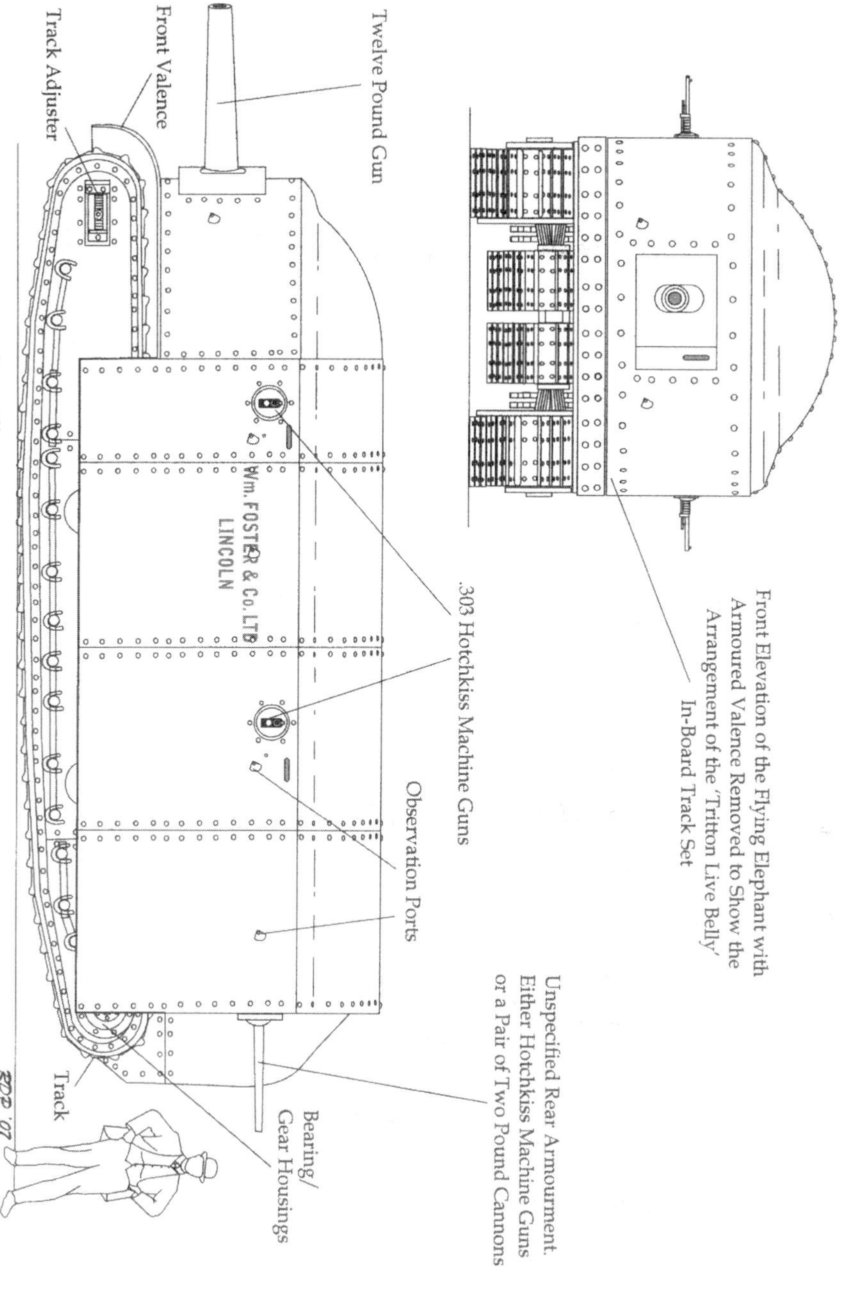

1916 William Foster and Co Ltd 'Flying Elephant' Prototype

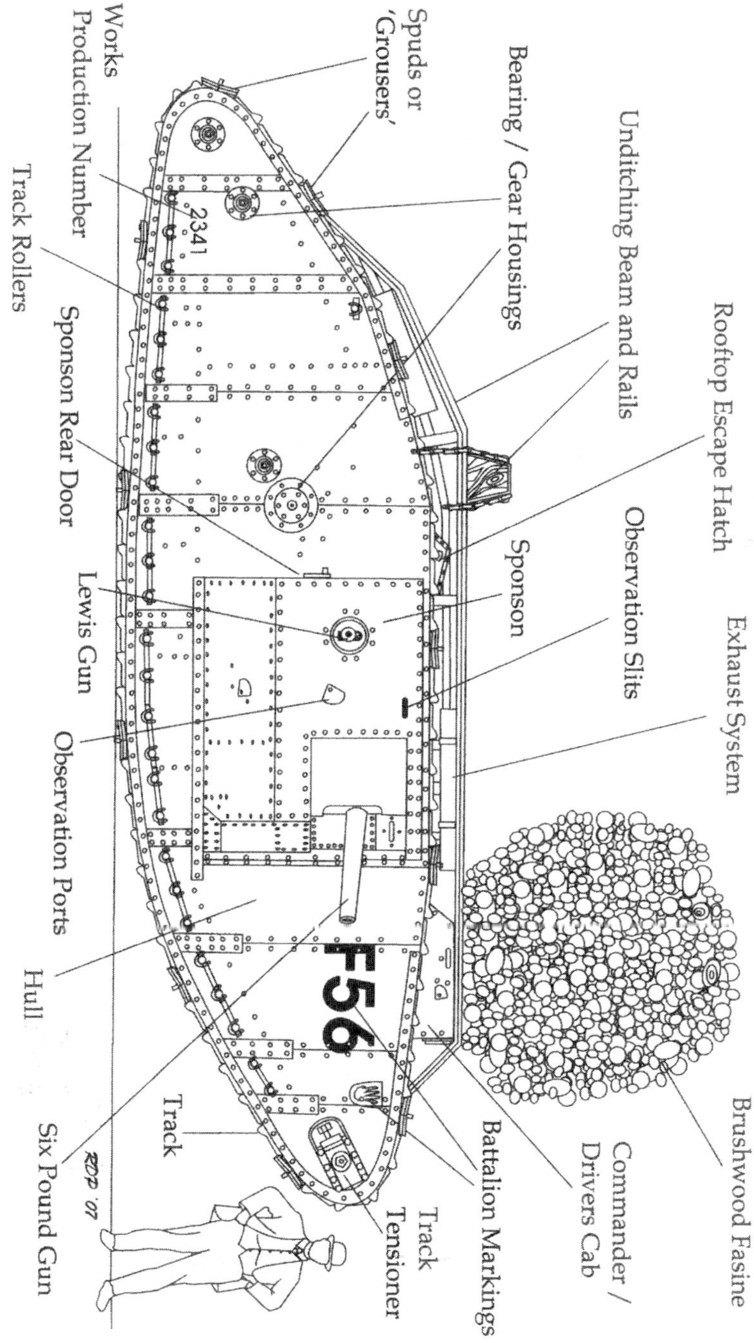

1917 William Foster and Co Ltd Male Mark IV Tank

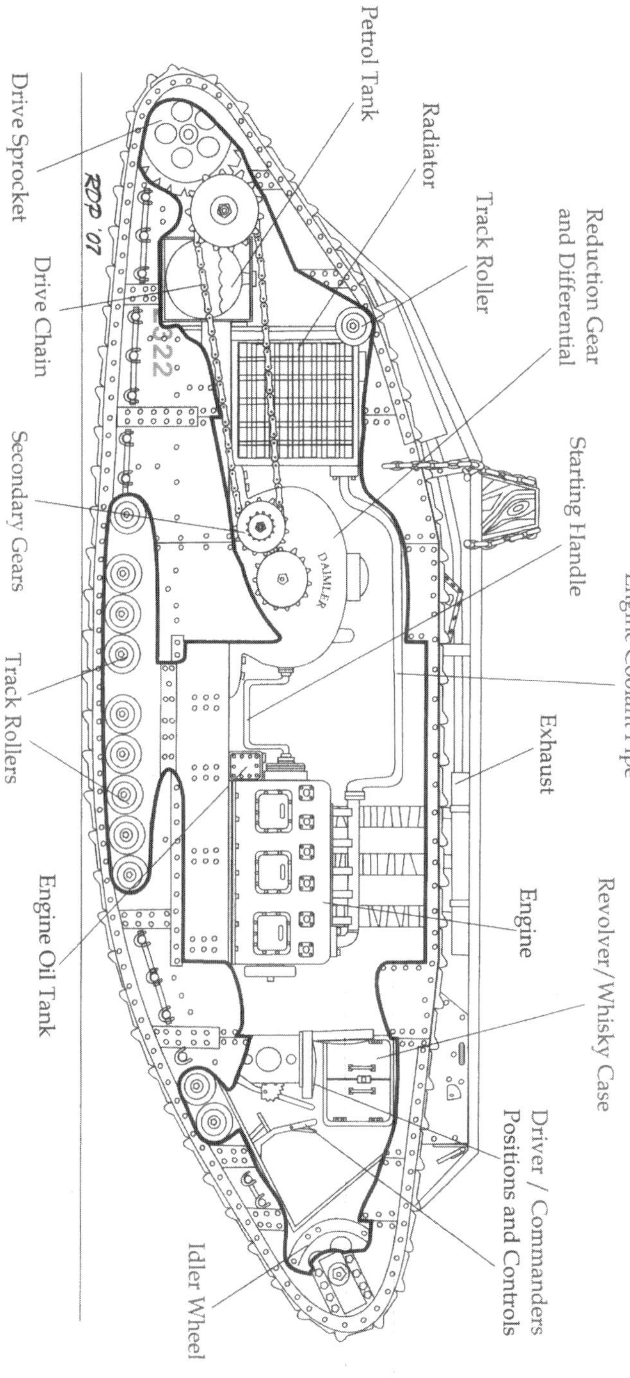

Cutaway Drawing Showing the Main Components and Internal Mechanical Arrangement of a 1917 Mark IV Tank

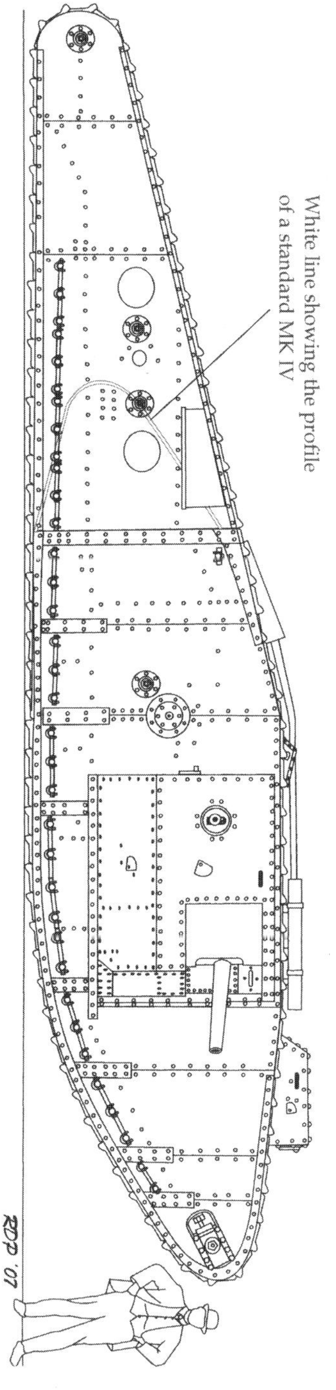

1917 William Foster and Co Ltd Mk IV with 'Tadpole Tail' Extension

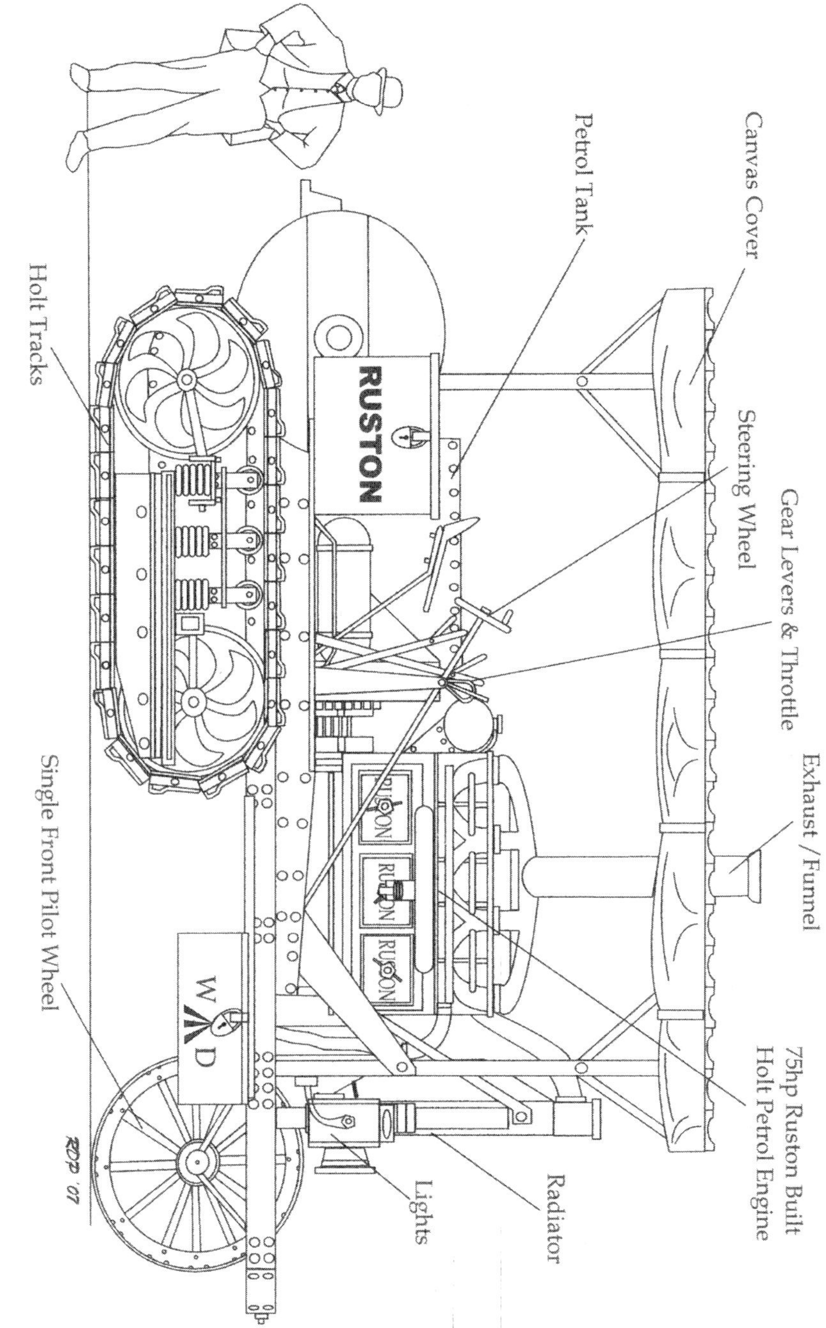

1917 Ruston & Hornsby Built Holt 75 hp Tractor

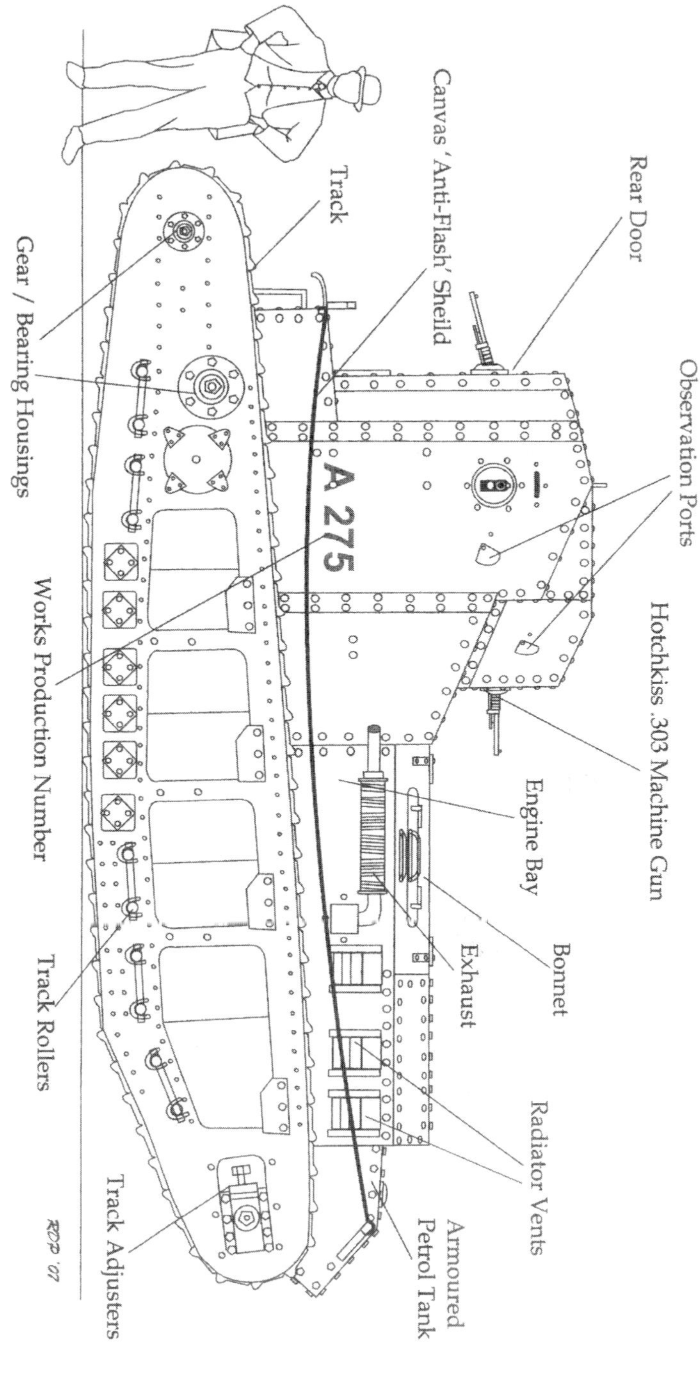

1917 William Foster and Co Ltd Medium A 'Whippet'

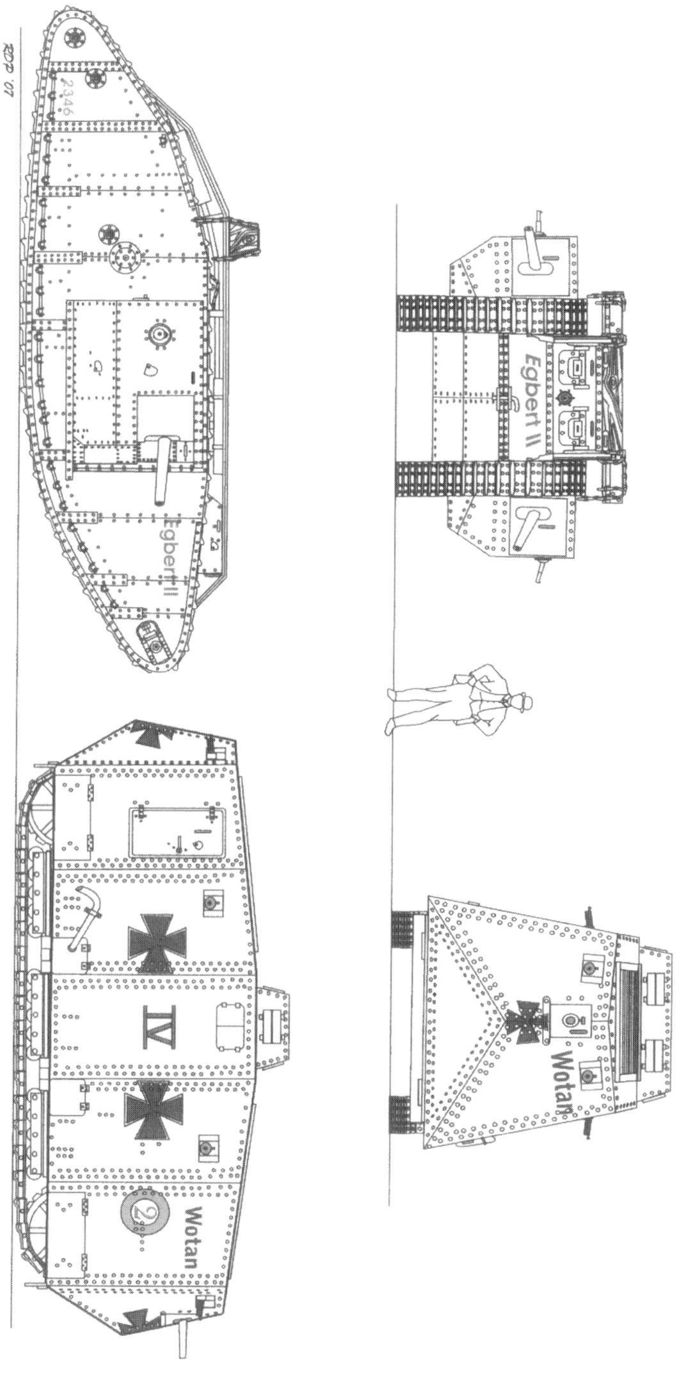

Front and Side Views Showing the Relative Sizes of a British Mk IV and a German A7V

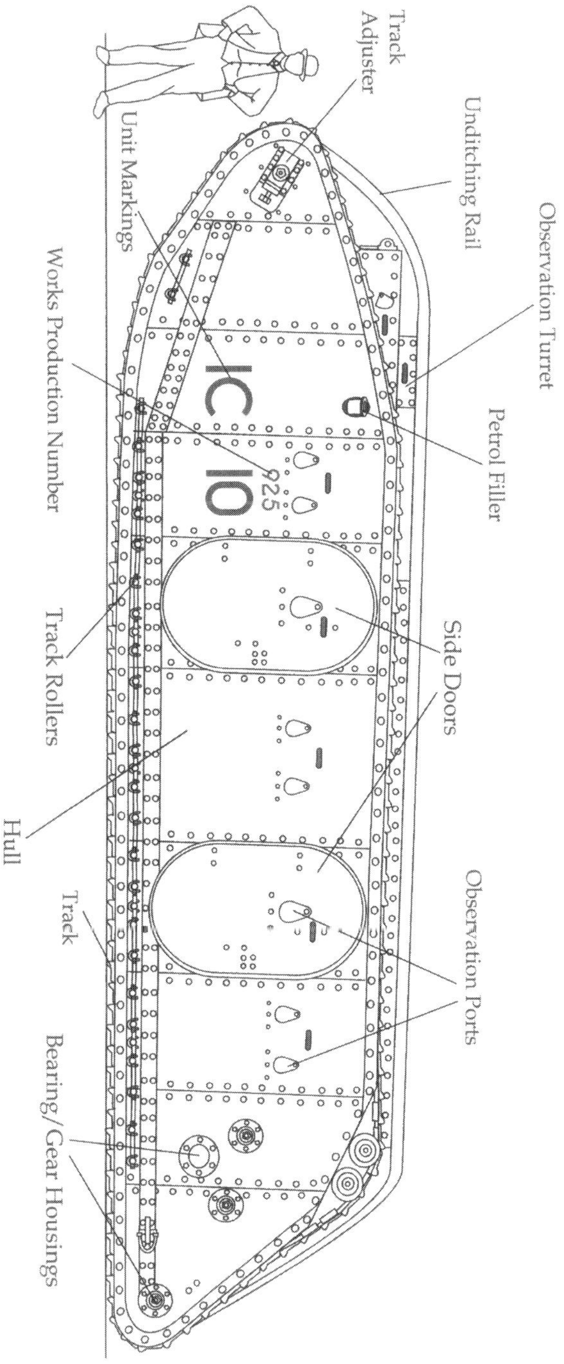

1918 Marshall Sons & Co Mark IX Supply Tank

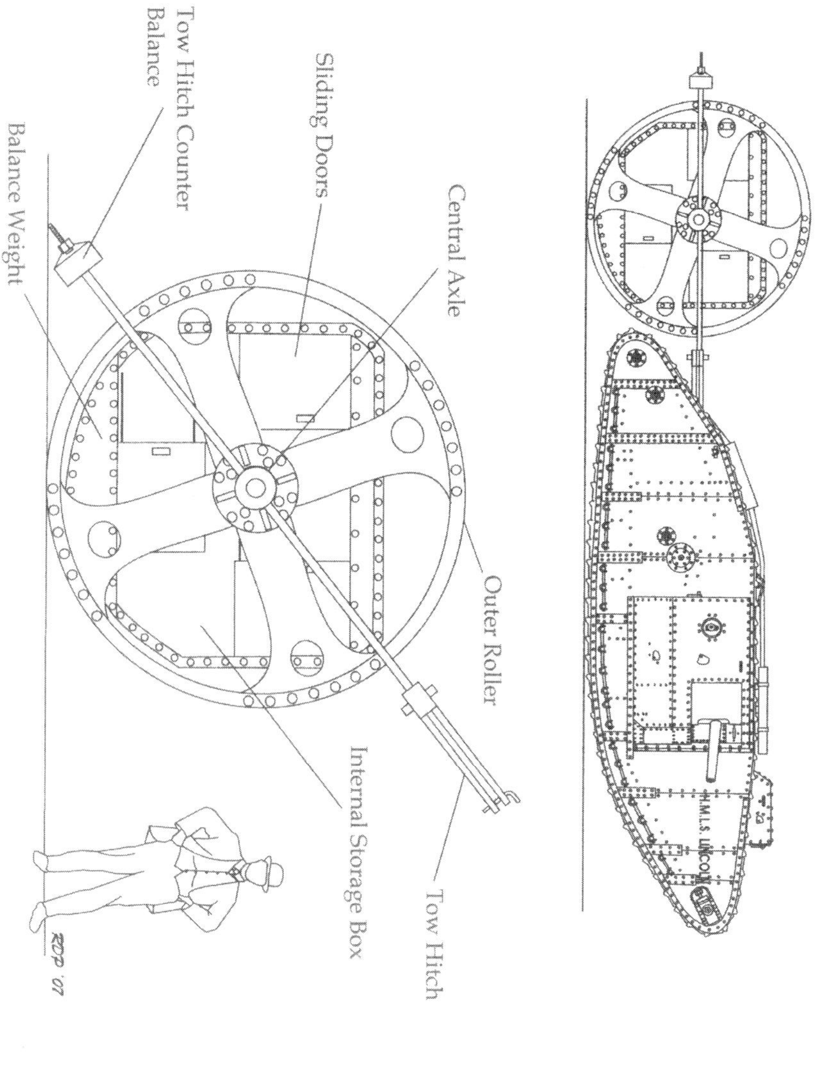

1918 Ruston & Proctor Stores Roller

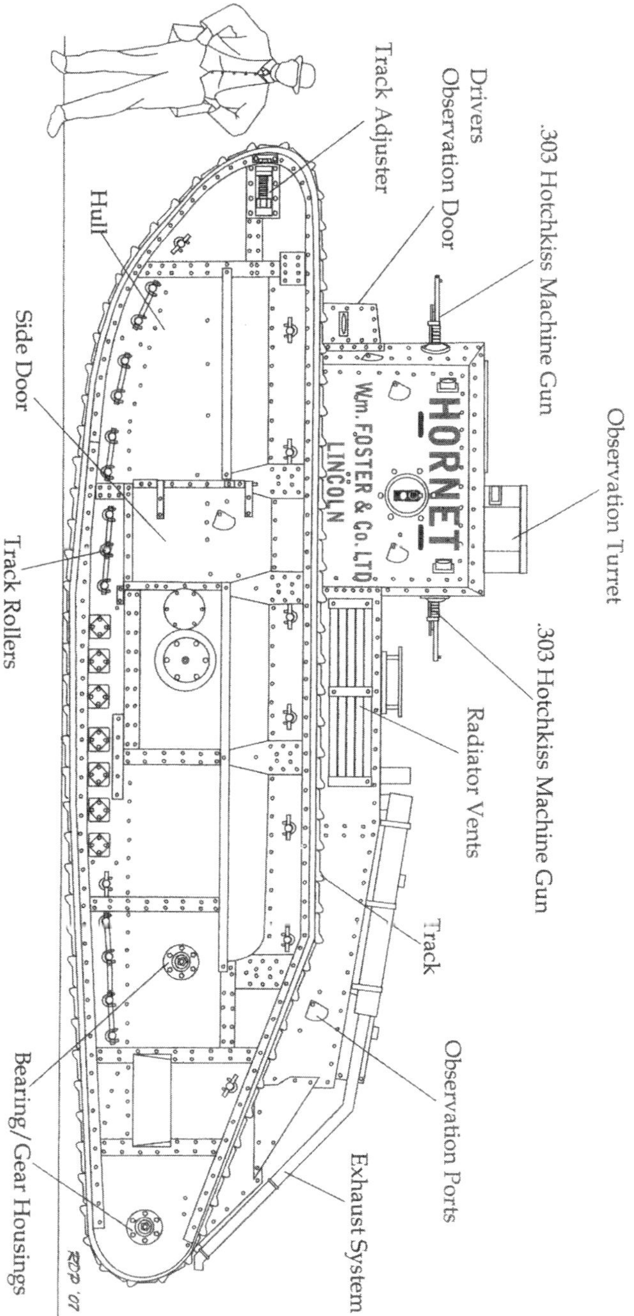

1918 William Foster and Co Ltd Medium C 'Hornet'

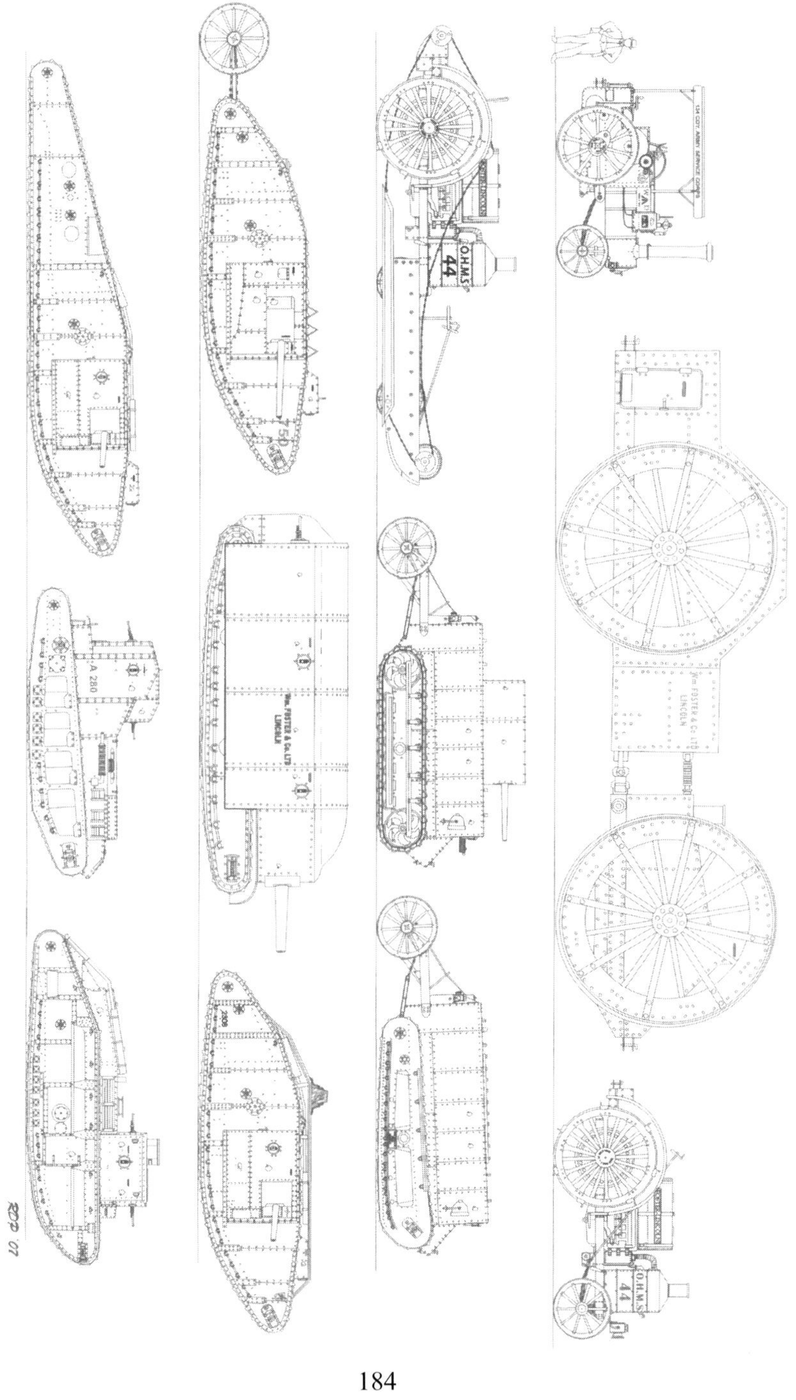

Size Comparison and Chronology of William Foster & Co Ltd Military Vehicles

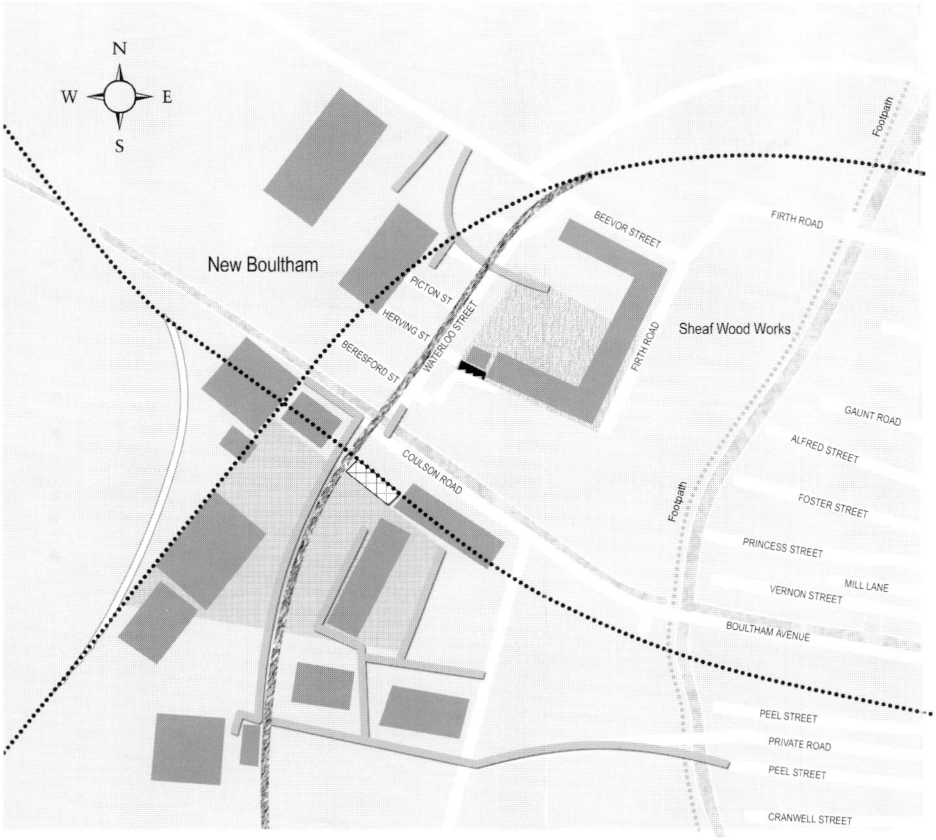

Map showing the locations of the former William Foster & Co Ltd. factory and testing ground with modern roads and buildings superimposed

(Based on a 1922 Lincoln Tourist Information Map)

BIBLIOGRAPHY

Recommended reading and acknowledgements

Anon.	1945	*A Short History of the Royal Tank Corps* Gale and Polden LTD, Aldershot
Bader P & Hart-Davis A.	1998	*More Local Heroes* Sutton Publishing LTD, Worcester
Beddows K. C & S Wheeler	1999	*Metro-Cammell, 150 years of Craftsmanship* K Beddows, C Wheeler, S Wheeler and Runpast Publishing
Boin W.R.	1997	*War Memoirs 1917-19* H.Karnac Books Ltd, London
Breyette T.W. Bender R.J.	2000	*Tank Killers, History of the Tank Destruction Badge* R. James Bender Publishing, California
Clark R.H.	1998	*Steam Engine Builders of Lincolnshire* Society for Lincolnshire History and Archaeology
Cooper B.	1974	*Tank Battles of World War I* Ian Allan Ltd, Shepperton
Fletcher D.	1984	*Landships. British Tanks of the first World War* HMSO, London
Fletcher D.	1991	*Mechanised Force, British Tanks Between the Wars* HMSO, London
Fletcher D.	1994	*Tanks and Trenches* Sutton Publishing, Stroud

Fletcher D.	2001	*The British Tanks 1915-19* Crowood Press, Ramsbury, Marlborough
Forty G. & A.	1988	*Bovington Tanks* Dorset Publishing Company, Sherborne, Dorset
Forty G.	1989	*Royal Tank Regiment, A Pictorial History* Guild Publishers
Anon.	2007	*Mechanical Maintenance of the Mark IV Tank* Friends of the Lincoln Tank, Lincoln
William Foster and Co LTD	1920	*The Tank. Its birth and development* Bemrose and Sons, Derby and London
Gibot J.L. Gorczynski P.	1999	*Following the Tanks, Cambrai* Imprimerie Centrale de l'Artois, Arras
Glanfield J.	2001	*The Devils Chariots* Sutton Publishing, Stroud
Guderian H.	1999	*Achtung Panzer!* Brockhampton Press
Lane M.R.	1997	*The Story of the Wellington Foundry, Lincoln* Unicorn Press, London
Liddell Hart B.H.	1934	*The History of the World War 1914-1918* Faber and Faber LTD, London
Liddell Hart B.H.	1959	*The Tanks. The history of the Royal Tank Regiment* *Volumes I and II* Cassell and Co LTD, London
The Lincoln Tank Group	1988	*Tank Papers* *Moorprint, Lincoln*
Macksey K.	1988	*Tank versus Tank* Guild Publishing

Mills D. (Editor)	1989	*History of Lincolnshire Volume XII; Twentieth Century - Lincolnshire* The History of Lincolnshire Committee, Lincoln
Mitchell F.	1987	*Tank Warfare* Spa Books, Stevenage
Parker C. & Walls J	2000	*Aircraft Made in Lincoln* Society for Lincolnshire History and Archaeology
Parker T.	1995	*The Tanks at Flers. Volumes I and II* Fairmile Books, Surrey
Peglar M.	1982	*The Tank Corps. Honours and Awards 1916-1919* Midland Medals, Birmingham.
Reynolds J.	1999	*Engines and Enterprise, The Life and Work of Sir Harry Ricardo* Sutton Publishing, Stroud
Rigby W.	1923	*The evolution of the Tank* Unpublished Article held by Lincoln City Archives
Ruston and Hornsby Ltd	1920	*Our Part in the Great War* Bemrose and Sons, Derby and London
Schneider W & Strasheim R	1990	*German Tanks in World War I. The A7V and Early Tank and development* Schiffer Publishing, Pennsylvania
Smithers A.J.	1986	*A New Excalibur* Book Club Associates
Stern A.	1919	*Tanks 1914-1918, The Logbook of a Pioneer* Richard Clay and Sons, London

Swinton E.D. 1932 *Eyewitness*
Hodder and Stoughton Limited, London

White B.T. 1978 *British Tank Markings and Names*
Squadron/Signal Publications, Michigan

Whitmore M. 1989 *Mephisto. A7V Sturmpanzerwagen 506*
Queensland Museum, Brisbane, Australia

Wilson A.G. 1986 *Walter Wilson: Portrait of an Engineer*
Gerald Duckworth and Co Ltd, London

Wright P. 2000 *Tank, the progress of a monstrous war machine*
Faber and Faber, London

Additional material and assistance was supplied by;
The Tank Museum, Bovington
Association Tank de Flesquieres, France
Musee de Royale de líarmee, Brussels
Museum of Lincolnshire Life, Lincoln
Bill Baker
Dennis Roy
Chris Gresham
Jean Letailles
Ray Hooley
Roy Wells
Tony Ablett
Lincoln City Library
Lincoln County Council
Lincolnshire Archives Office
The Lincolnshire Echo Newspaper
The Lincolnshire Leader Newspaper
The Lincolnshire Gazette Newspaper
Newark and Sherwood District Council